图书在版编目(CIP)数据

遇见新古典. 2: 汉英对照 / 胡延利, 崔淼主编. —大连: 大连理工大学出版社, 2013.6
ISBN 978-7-5611-7834-8

Ⅰ.①遇… Ⅱ.①胡… ②崔… Ⅲ.①住宅—建筑设计—作品集—世界—现代 Ⅳ.①TU241

中国版本图书馆CIP数据核字(2013)第096273号

出版发行：大连理工大学出版社
（地址：大连市软件园路 80 号　邮编：116023）
印　　刷：深圳市彩美印刷有限公司
幅面尺寸：235mm×305mm
印　　张：19.25
出版时间：2013 年 6 月第 1 版
印刷时间：2013 年 6 月第 1 次印刷
总 策 划：周群卫
责任编辑：张昕焱
封面设计：周群卫 王志峰
责任校对：石雅新

书　　号：ISBN 978-7-5611-7834-8
定　　价：328.00 元

发　行：0411-84708842
传　真：0411-84701466
E-mail: 12282980@qq.com
URL: http://www.dutp.cn

遇见新古典

Engaging with the Classics: Collection of New Traditional Residential Masterworks

2

胡延利 崔淼 主编

卢晓娟、王思锐、时跃、杨宇芳、连飞飞 译

大连理工大学出版社

内容介绍
Introduction

《遇见新古典：新式传统住宅作品集 2》延续了第一册的编辑理念，收录了传统住宅建筑杰作，旨在将更多致力于传统建筑的知名建筑师及其优秀的建筑作品呈现给专业读者，如 Andrew Skurman 建筑师事务所、Curtis & Windham 建筑师事务所、David Jones 建筑师事务所、Derrick 建筑事务所、Domani 建筑与规划事务所、Fairfax & Sammons 建筑设计公司、Granoff 建筑师事务所，Landry 设计集团以及 Oatman 建筑设计公司。

该书向读者展示了新式传统住宅设计的最佳案例，它们将以一种更富活力与创新性的方式来诠释传统建筑，并适应技术与生活方式等现代需求。建筑师的学识、能力、思想、直觉、想象力以及他们对人文与环境的关注所共同迸发出的潜在力量将为传统建筑赋予一层崭新的意义。

This new collection on traditional residential architecture continues the idea of the first volume, aiming at introducing worldwide renowned architects who practice in the classical tradition and their excellent works to professional readers. The book features the masterworks from famous classical architects like Andrew Skurman Architects, Curtis & Windham Architects, David Jones Architects, Derrick Architecture, Domani Architecture + Planning Inc., Fairfax & Sammons Architects P.C., Granoff Architects, Landry Design Group and Oatman Architects, Inc..

The book will provide the finest examples of new classical residential designs, which interpret the traditional principles in a vital, innovative and ingenious manner and adapt to contemporary needs of technology and lifestyles. The potential power inspired by the combination of the architects' knowledge, skill, thought, intuition and strength of imagination, as well as their attention to humanist and environment, endows the traditional architecture with new meaning.

目录
Contents

Andrew Skurman 建筑师事务所

Andrew Skurman Architects

建筑师 Andrew Skurman 于 1992 年在旧金山成立了他的公司。作为 Andrew Skurman 建筑师事务所的负责人和所有者，他总是完美且合理地根据甲方的要求和愿望来设计出极其精致的住宅。他善于表现古典建筑的优雅与精致，并能够对法国、乔治王时代和地中海风格进行完美的诠释。

Skurman 已经荣幸地由法国文化部授予艺术与文学勋章。他是著名的古典建筑与传统美国建筑研究所顾问委员会的成员。目前，他在旧金山秋季古董展担任创意总监。

迄今为止，Skurman 在加利福尼亚北部和南部、纽约、内华达、法国和中国都有设计项目。他拥有加利福尼亚和纽约的建筑师资格证书，并在旧金山和巴黎两地居住。

Skurman 于 1976 年获得纽约库珀联合学院的建筑学学士学位，随后在一些世界上最负盛名的建筑公司工作。从 1976 年至 1980 年，他在贝聿铭纽约事务所做学徒，这是他设计生涯的开始。然后，他于 1980 年至 1987 年在 SOM 建筑工程公司旧金山办事处担任高级助理。1987 年至 1992 年，他又分别在 Gensler & Associates 公司的旧金山和洛杉矶工作室担任总监一职。

由 Andrew Skurman 建筑师事务所设计的住宅已在众多出版物上刊登过，如《建筑文摘》《住宅与花园》《南方风格》《法国住宅》《纽约时报》《西方室内设计》《加利福尼亚住宅》《加利福尼亚住宅与设计》《C 杂志》《旧金山杂志》《老房子》《罗博报告》《奢华室内与设计》《完美住宅》《传统住宅》以及《贵族设计》。该公司的设计作品也被收录在凯瑟琳•马森所著的《纳帕山谷风格》（Rizzoli，2003 年）一书中，以及黛安•多兰斯•塞克斯所著的《旧金山风格》（编年史书，2004 年）一书中。

Architect Andrew Skurman founded his firm in San Francisco in 1992. As principal and owner of Andrew Skurman Architects, he focuses on superbly crafted custom houses that are perfectly and logically planned to the specific requirements and wishes of his clients. His expertise lies in the elegant and refined expression of Classical architecture and the interpretation of French, Georgian, and Mediterranean styles.

Skurman has received the honor of being named a Chevalier of Arts & Letters by the Minister of Culture of France. He is an appointed member of the prestigious Council of Advisors of the National Institute of Classical Architecture and Classical America. He currently serves as Creative Director of the San Francisco Fall Antique Show.

Skurman is currently designing projects in Northern and Southern California, New York, Nevada, France and China. He holds architectural licenses in California and in New York, and resides in both San Francisco and Paris.

Skurman received his Bachelor Degree of Architecture in 1976 from Cooper Union in New York City and subsequently worked at some of the most prestigious architectural firms in the world. He began his design career apprenticing with the New York firm of I.M. Pei & Partners from 1976 to 1980. He then worked in the San Francisco office of Skidmore, Owings & Merrill as a Senior Associate from 1980 to 1987 and then as a Studio Director at Gensler and Associates in both San Francisco and Los Angeles until 1992.

Homes designed by Andrew Skurman Architects have been featured in numerous publications such as *Architectural Digest, House & Garden, Southern Accents, Maison Française, The New York Times Magazine, Western Interiors, California Homes, California Home & Design, C Magazine, San Francisco Magazine, This Old House, The Robb Report, Luxe. Interiors + Design, House Beautiful, Traditional Home*, and *Gentry Design*. Work by the firm is also included in the books *Napa Valley Style* (Rizzoli, 2003) by Kathryn Masson and *San Francisco Style* (Chronicle Books, 2004) by Diane Dorrans Saeks.

作　品

Selected Work

乡村住宅
A Country Residence

乡村住宅
A Country Residence

美国
USA

之前的一位甲方的父母来找建筑师设计他们的新住所。他们希望在退休以后拥有一座宜居的、交通便捷的住宅，并且想让该住宅的所有主要空间都设置在一层。在建筑师办公室中的图书室会面之后，他们对凡尔赛的两个楼阁表示出异常的喜爱，进而决定他们的房子要有非常明显的法国风情。

为了获得这种18世纪晚期的建筑风格，建筑师们为这座住宅增设了一个巨大的折线形屋顶和几个拱形天窗。由于该住宅只有一层，因此他们在设计室内空间时就拥有很大的自由发挥空间，而不会因为没有上层或者过浅的屋顶而影响设计。四个圆形门廊都安装了圆形屋顶天窗，住宅的其他空间也设有这种天窗，从而使得内部空间可以充满自然光线。

与许多其他的经典法国建筑类似，这座U形住宅也设有外墙角。U形布局是建筑师个人最喜欢的格局，因为它使每个房间都充满自然光线。这种形式促使空间纵横交错，使人们在住宅不同的体量中行走时，可以充分体验到空间变换所带来的新鲜感。

该住宅拥有两个侧翼，其中一个里面设有主人套房，另一个里面设有厨房、早餐室和书房。中间部分是公共空间。一个大型的采光天窗照亮了入口大厅和主厅，主厅中精心安装了三个大窗户，透过它们可以看到室外的游泳池，从而为主厅室内提供了一个迷人的背景。建筑师尽力将该住宅中的每个细节与整体联系到一起，以使其看起来简单却不失优雅。

The parents of one of architect's previous clients came to him to design their new residence. Looking towards their retirement, they wanted an easily livable and accessible house, which created the programmatic requirement of having all the major spaces located on the ground level. After meeting in his office's library, they decided on a distinctly French character upon falling in love with two small pavilions in Versailles.

In order to obtain the necessary proportions required of this late eighteenth century style, a large mansard roof with segmental arched dormers was added. Due to the one story nature of the house, this allowed for greater freedom in the sculpting of the interior spaces as there was no upper level or shallow roof holding back the design. Each of the four circular vestibules is topped with a domed lay light; other lay lights are present throughout the house, allowing for natural lighting of the interior spaces.

Quoins, similar to those of many classical French buildings, bookend the corners of the U-shaped building. The U- shaped layout is a personal favorite of his, as it allows natural light into every room. This form also forces spaces to fall on axis with one another in enfilade, creating an exciting feeling of movement as one walks through the various volumes.

One wing of the house is devoted to the master suite, while the other contains the kitchen, breakfast room and his study. The central portion contains the public spaces. A large lay light illuminates the entrance hall on axis with the great hall, while three large windows provide a stunning backdrop with a view of the pool for the exquisitely paneled interiors of the great hall. This house achieves a simple elegance through a conscious effort to relate every detail back to the whole.

后立面　**Rear Elevation**

正立面　**Front Elevation**

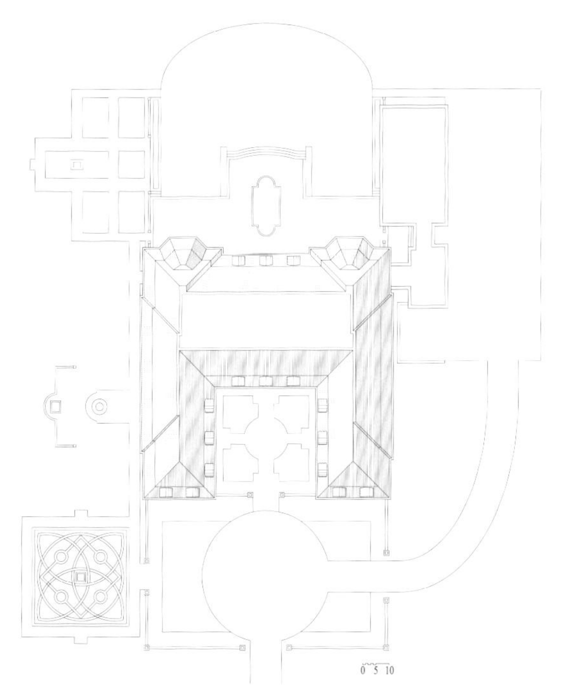

一层平面图 First Floor Plan

$\frac{1}{16}$"=1'-0"

VILLE VENETE
PARIS IN BLOOM

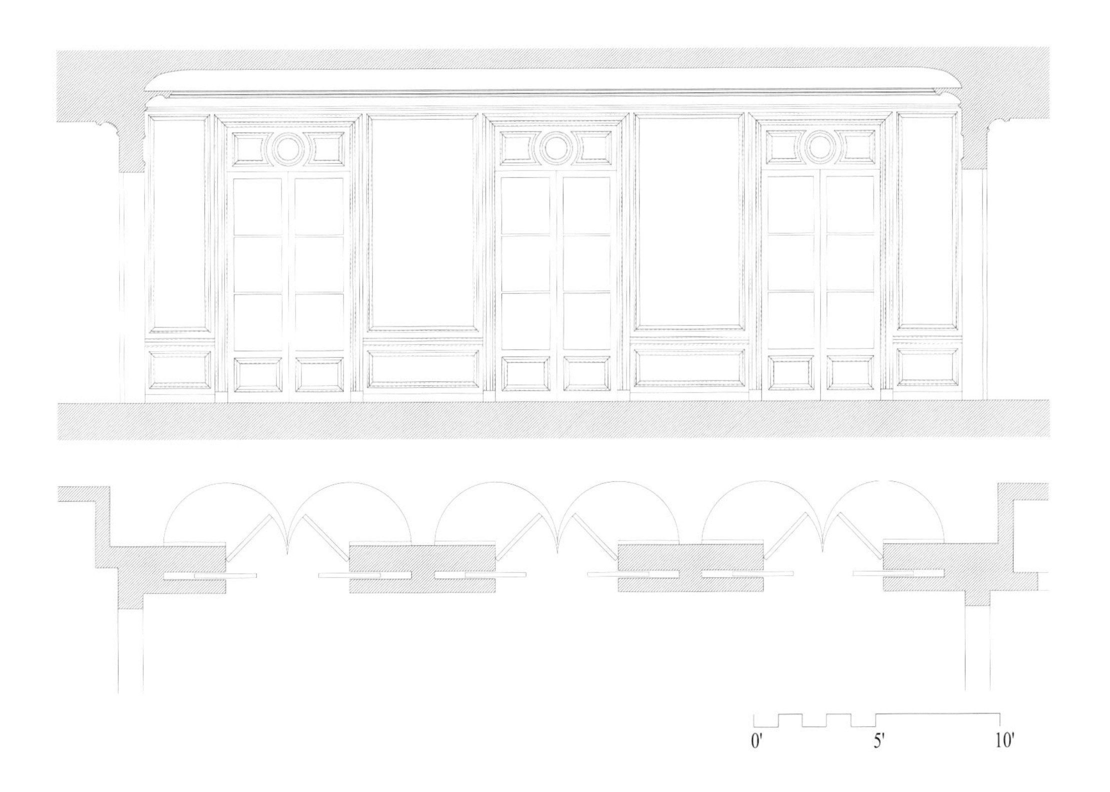
0'
5'
10'

THE GIVENCHY
Le musée des

Curtis & Windham 建筑师事务所

Curtis & Windham Architects

Curtis & Windham 建筑师事务所于 1992 年在德克萨斯州的休斯敦成立，是一家提供建筑设计、景观设计和室内设计服务的建筑和景观类建筑公司。从公司负责人比尔·柯蒂斯和拉塞尔·温德姆合作伊始，他们就在建造传统建筑方面产生了共同的兴趣，传统建筑将对建筑历史和环境的深深敬意和传统经典细部的研究知识结合起来。柯蒂斯先生和温德姆先生在他们的建筑师事务所在休斯敦建立之前，就在他们的故土德克萨斯州以外的地方开始了他们的事业。在华盛顿和伦敦花费的时间，使两人都对建筑历史、城市环境和传统建筑的用途萌生了尊重之情。2005 年，该事务所的建筑实践拓展到景观建筑领域。他们的作品展示了事务所对大量传统设计方法的掌握和对符合现代生活方式的当地建筑风格的理解。该公司在许多地方都有设计作品，其中包括纽约长岛的木瓦风格建筑、美国西部的几个大型农场、佛罗里达海边的滨海小屋，以及代表了德克萨斯州休斯敦市一系列历史悠久社区的建筑风格的住宅。设计和建造过程的高水准实施是 Curtis & Windham 建筑师事务所成功的关键。他们的作品在大量出版物中出现，被许多当地和国际大赛所认可。

Founded in 1992 in Houston, Texas, Curtis & Windham Architects began as a general practice of architecture evolving to later include landscape architecture and interior design. Since the beginning, principals Bill Curtis and Russell Windham have shared a mutual interest in making buildings that combine a deep respect for architectural history with a well-studied knowledge of traditional and classical principals and detailing. The eclectic nature of their work demonstrates a thorough command of traditional architecture in the broadest sense, while their ability to create a vernacular architecture that can gracefully accommodate a modern lifestyle evidences both a commitment to their clients' needs and the continued relevance of traditional and classical design. The landscape studio of Curtis & Windham, established in 2005, designs and coordinates work both independently and in close collaboration with the architecture studios. The firm has completed work in a wide variety of locations, including an estate in New York ' s Long Island, small and large-scale ranch properties in the American West, beach cottages in Seaside, Florida, and primary residences in Houston's most desirable historic neighborhoods. The highest standards of execution in both design and construction are central tenets to Curtis & Windham ' s practice. Their work has appeared in numerous publications and has been recognized by both local and national awards.

作　品

Selected Work

Willowick 住宅
Willowick Residence

因伍德泳池亭
Inwood Pool Pavilion

Knollwood 公寓和花园
Knollwood Residence & Gardens

沉睡谷住宅花园
Sleepy Hollow Residence & Gardens

橡树河大道的花园凉亭
River Oaks Boulevard Garden Pavilion

Willowick 住宅
Willowick Residence

德克萨斯州休斯敦
Houston, Texas

该住宅占据了休斯敦橡树河住宅区内的水牛河河岸一处僻静的 V 形地块，该地区的典型特征是成熟的树木形成的如森林般的华盖和广阔的土地。这座住宅建在一条弯曲车道的尽头，沿路要通过一片类似于公园的广阔区域，周边遍布高大的橡树和松树。住宅周边是一家大型的碎石汽车旅馆。该住宅具有休闲的特质，灵感来自于英国建筑传统，而同时又有明显的美式风格。建筑师参照了大不列颠萨里郡庄园住宅的传统，特别是埃德温•鲁琴斯爵士的作品。项目的功能空间围绕着由楼梯大厅和主入口构成的中心体量布置，两侧是不同尺寸和等级的辅助建筑。多边形的平面布局将建筑场地分成了多处室外空间，其中包括各种各样的花园，这些花园由鲁琴斯爵士的专业搭档，英国著名花园设计师格特鲁德•杰基尔设计。在住宅的后面可以看到水牛河南岸上森林茂密的斜坡。一系列干式堆叠的挡土墙、青石板阶梯和延伸至河岸的小径都增加了住宅与河岸之间的联系。杰基尔 / 鲁琴斯在艺术上的合作反映了室外正式和非正式区域的结合，例如，露台、大门、格子凉亭和石台阶，这些元素都在种满各种当地植物的花园的作用下变得柔和。项目体量的变化给人一种住宅被逐渐拓展的感觉，以适应历代发展对功能要求的改变。这种持久的感觉通过室外材料的颜色得到进一步的加强：垂直的柏木护墙板、厚重的杉木斜屋顶以及由长石切割成的石材。各式各样的双层高凸窗、入口门廊、带雕刻图案的石砌门框和实木阳台设计均借鉴了鲁琴斯的作品，但室内外均采用了当地材料，例如用于前烟草仓库的柏木和大胡桃木地板，既符合了英国建筑风格，也符合德克萨斯本地模式。

摄影：Hester +Hardaway Photographers

This residence occupies a secluded, wedge-shaped lot on the bank of Buffalo Bayou in a section of Houston's River Oaks neighborhood typified by a forest-like canopy of mature trees and generous lots. The house is found at the end of a winding drive through a park-like expanse of tall oaks and pine, and is organized around a large gravel motor court. The house is rambling in character and inspired by the English architectural style, while distinctly American in attitude. The architects looked to the tradition of manor houses found in Great Britain's Surrey County, specifically to the work of Sir Edwin Landseer Lutyens. The program is organized around a large central volume defined by the stair hall and main entry with dependent wings of varying scale and hierarchy. The angular plan shape serves to divide the site into several exterior spaces, including various gardens designed in the tradition of the great English garden designer and professional collaborator of Lutyens: Gertrude Jekyll. The back of the house looks out onto a densely wooded slope of the south bank of Buffalo Bayou. A network of dry-stack retaining walls, bluestone steps, and pathways cut into the bank enhance the relationship of the house to the bayou. The Jekyll/Lutyens aesthetic partnership is echoed in the composition of formal and informal exterior spaces, exemplified by terraces, gates, trellises, and stone steps that are softened by varied gardens of native plants. Variation in massing gives the sense that the house has been expanded to accommodate changing functions over generations. The sense of permanence is reinforced by the palette of the exterior materials; sinker-cypress siding, thick cedar shake roof and ashlar-cut stone. The various double-height window bays, entry porches, carved stone door surrounds, and substantial wood balconies quote from the work of Lutyens but use of local materials both on the interior and exterior, such as sinker-cypress and pecan-wood floors from a former tobacco barn adapt the English style to its Texas locale.

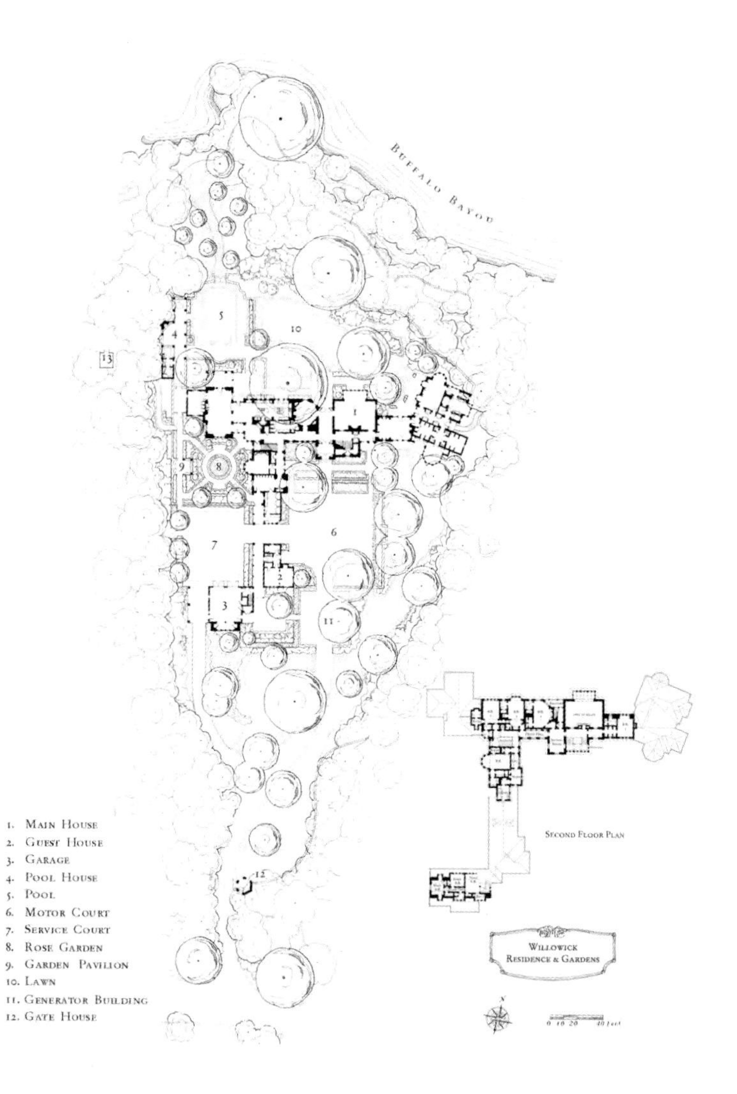

因伍德泳池亭
Inwood Pool Pavilion

德克萨斯州休斯敦
Houston, Texas

作为休斯敦建筑师约翰·斯托布于1963年设计的一座著名的历史住宅的附属建筑，这座由砖和石灰石建成的亭子成为这座德克萨斯州休斯敦市的住宅后庭中的建筑焦点。这座亭子在建筑场地上的布局类似于一座橘园，与横穿一片草坪和一条栽种橡树的大道的原有住宅相垂直。亭子由带有风化的油漆罩面的砖所覆盖，材料的颜色与住宅的颜色一致。亭子的室内主要为木质装饰，而外部装饰为雕刻石灰石，暗示了其更多独特的功能。亭子内规划了一处大型娱乐空间，娱乐空间沿着容纳大量服务功能的厨房和储藏区域设置，位于一个通过精心设计的比例和细部来遮盖其规模的一个结构之内。通过应用具有英国摄政时期风格的平整而纤细的装饰，亭子立面与已有住宅的正立面相得益彰。层状的装饰展示了建筑的构造，砖拱由曲形的石灰石圆拱和拱心石支撑。外部圆拱的槽纹细部和壁缘上的希腊回纹都暗示了室内的几何设计。三个拱形洞口主宰了室外顺序，而在室内，固定高处穹窿的一块石膏檐板和槽型顶棚界定了宴会厅的轮廓。在亭子仍然真实地保持英国摄政时期风格的比例的同时，更多受到限制的外观被更加丰富的室内所取代：黑白相间的大理石镶嵌地板、带有花纹细部的石膏顶棚以及带有几何图案的、装有镜子的浴室，都唤起了艺术装饰时期的记忆。最后，该亭子结合了两种设计思路：一座附属的建筑仍然作为已有住宅的附属物，同时在不加限制室内规模和装饰的情况下体现豪华娱乐设施的功能。

摄影：Hester +Hardaway Photographers

Built to accompany a historic residence designed by Houston architect John Staub in 1936, this brick and limestone pavilion is the architectural focus in the backyard of this Houston, Texas property. The placement of the pavilion on the site is analogous to an orangery, located perpendicular to the existing house across a lawn and allée of live oak trees. Clad in brick with a weathered paint finish, the material palette of the pavilion follows that of the house. Where accents are rendered in wood at the house however, the exterior ornament of carved limestone used on the pavilion hints at the building's more distinctive function. The pavilion's program consists of a large entertainment space along with generous service kitchen and storage areas, housed inside a structure that masks its size with carefully designed proportions and detail. By applying the flattened and attenuated ornament of the Regency style, the pavilion facade complements the facing elevation of the existing house. The layered ornament suggests the tectonics of the building, where brick arches are supported by carved limestone arches and keystones. Fluted detail at the exterior arches and a Greek key at the frieze hint at the geometric design of the interiors. Three arched openings dominate the exterior order, while on the inside, a plaster cornice anchoring a lofty cove and tray ceiling define the ballroom. While still remaining true to the proportions of the Regency style, the more restrained exterior gives way to a more exuberant interior: black-and-white marble inlay floors, plaster ceiling with floral detail, and mirrored bath with geometric motif is evocative of the Art Deco period. Ultimately, the pavilion integrates two design forces; the duty of an accessory building to remain architecturally subordinate to the existing house while expressing with less restrained interior scale and ornament, the function of grand entertainment.

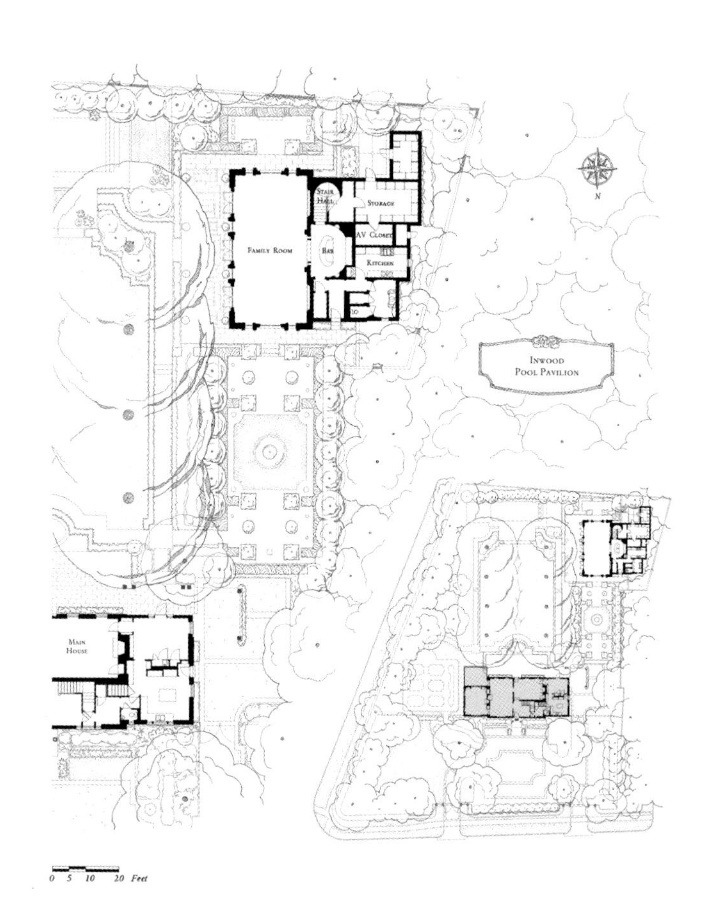

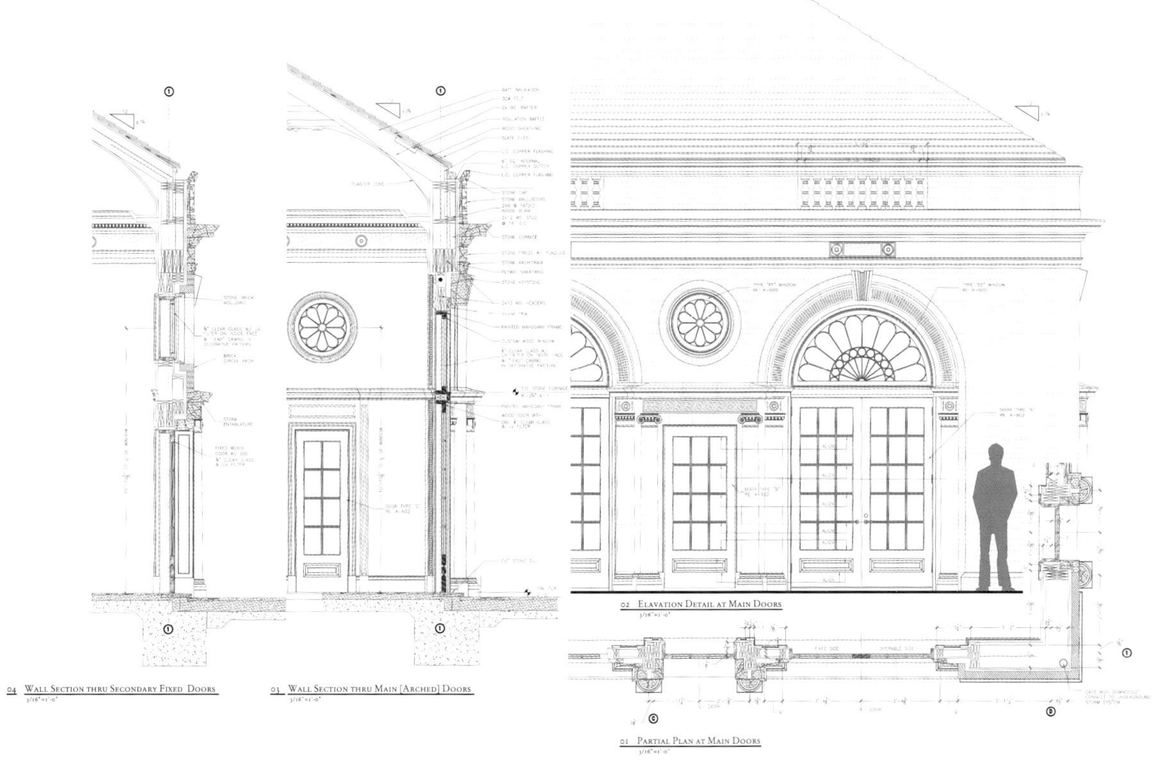

04 Wall Section thru Secondary Fixed Doors

03 Wall Section thru Main [Arched] Doors

02 Elevation Detail at Main Doors

01 Partial Plan at Main Doors

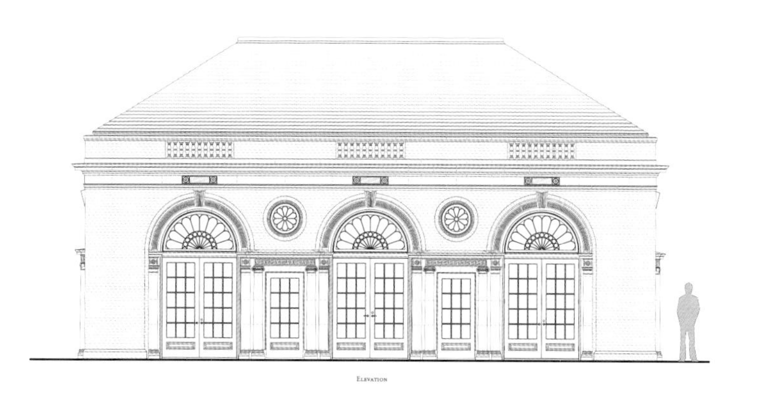

Elevation

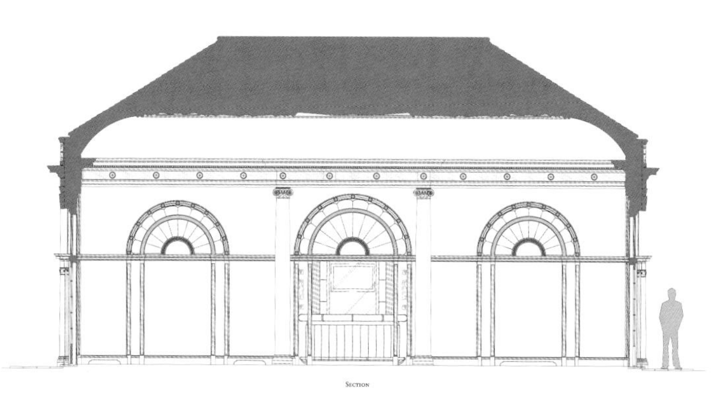

Section

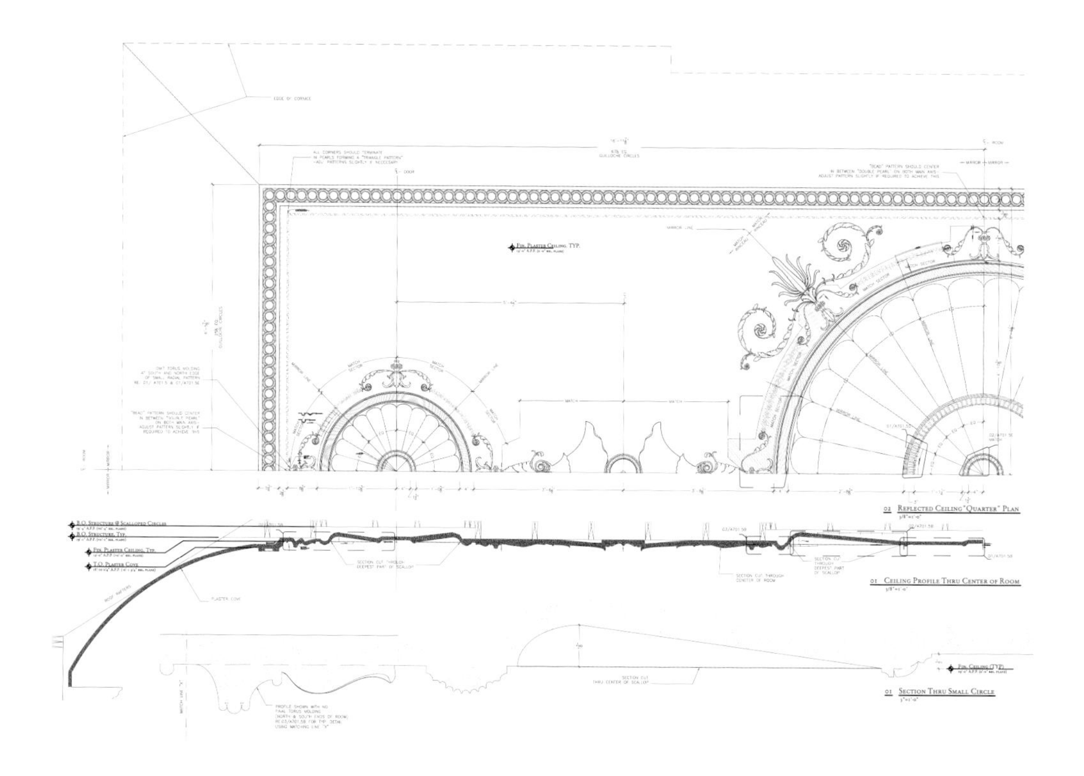

Inwood Pool Pavilion

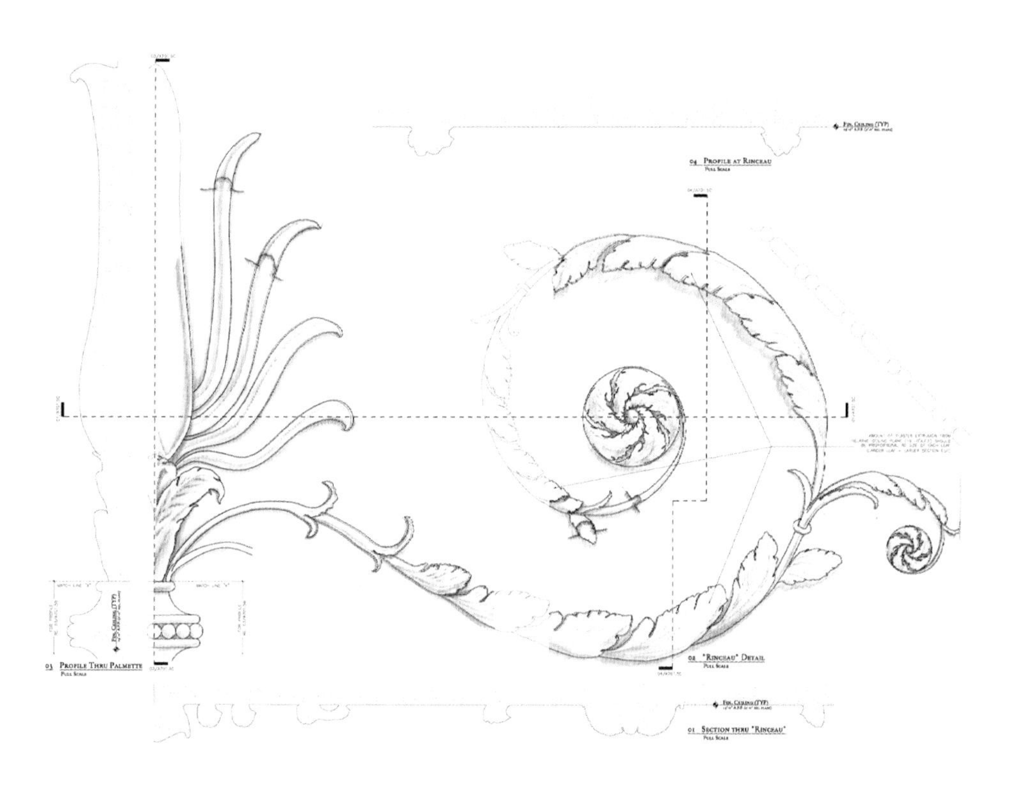

Inwood Pool Pavilion

Knollwood 公寓和花园
Knollwood Residence & Gardens

德克萨斯州休斯敦
Houston, Texas

这座位于休斯敦的公寓以及该公寓和花园之间的布局关系的设计灵感来自于建筑师、景观设计师查尔斯•A•普拉特的作品。如何处理房子和给定场地之间的关系，建筑师对以下几个方面进行了细致周密的思考：使建筑面向边道而非主道，以便留出宽敞的前院修筑一条规整的入户通道。同时，使用石子在屋后空间铺设一条环形机动车道，直通车库。建筑的朝向、体长和短轴的设计充分利用了花园中紧凑的景观。花园中还建造了几处外部空间，其中最主要的部分就是屋前宽敞的椭圆形草坪、北面的角树丛以及泳池露台。本设计既关注了住宅与景观的布局关系，同时也考虑到了建筑墙体与开口之间的整体构造和比例问题，这些都体现了普拉特建筑作品的精神。建筑主体从平面图和立面图来看皆均匀对称，里面的房间中规中矩，北面有一个较矮的侧翼建筑和含有服务区的通道。内部房间的布局高效合理，交通流线主要是在各个房间之间实现的。整个建筑体量生动地扎根在地面上，混凝土做成的泻水台从墙体表面探了出来。灰泥粉饰的建筑外壁没有着色，给人一种粗砾砂酸洗后形成的中性色调的感觉。与此相应，外部色调的搭配也同样朴实无华：混凝土浇铸的特色、再生粘土瓦屋顶、沉积柏木制成的格子墙以及被漆成蓝色的实木窗和百叶窗，这些都凸显了严谨内敛的外部装饰风格。

摄影：Hester +Hardaway Photographers

This Houston residence and its relationship to the garden were inspired by the work of architect and landscape designer Charles A. Platt. The relationship of the house to the given site was carefully considered in several ways: a formal "frontal" approach to the house is provided through a generous front yard by orienting the house to the side street rather than the address street. In turn, the space behind the house is occupied by a circular gravel-paved motor court, with access to the garage. The orientation and major and minor axis of the house are designed to take advantage of tightly designed views of the garden where several exterior rooms have been created, the most significant of which are the large oval lawn in the front, a hornbeam bosque to the north, and the pool terrace. The attention given to the relationship of house to landscape as well as the massing and proportion of wall to openings reflect the spirit of Platt's work. The main block of the house containing the formal rooms is symmetrical in plan and elevation, with a lower wing and hyphen containing service areas to the north. The rooms inside are efficiently organized, and circulation occurs mostly from room to room. The volume of the house is visually anchored to the site with a concrete water table, projected from the plane of the wall surface. The stucco exterior is made without colored tinting, and derives its neutral tone from torpedo sand exposed in an acid wash. The balance of the exterior palette is similarly humble: cast concrete accents, reclaimed clay tile roof, sinker-cypress trellis and blue-painted wood windows and shutters underscore the restrained exterior ornament.

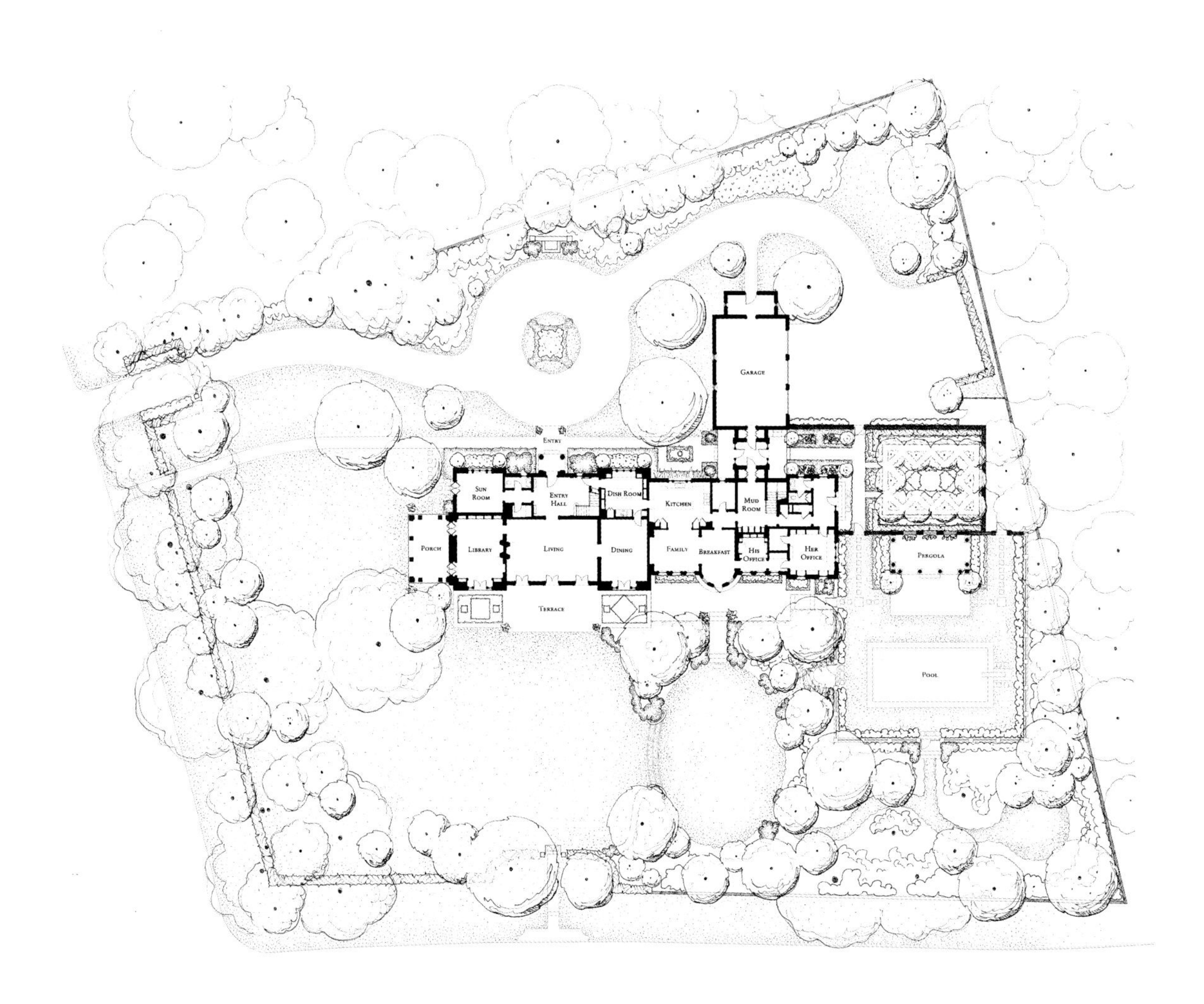

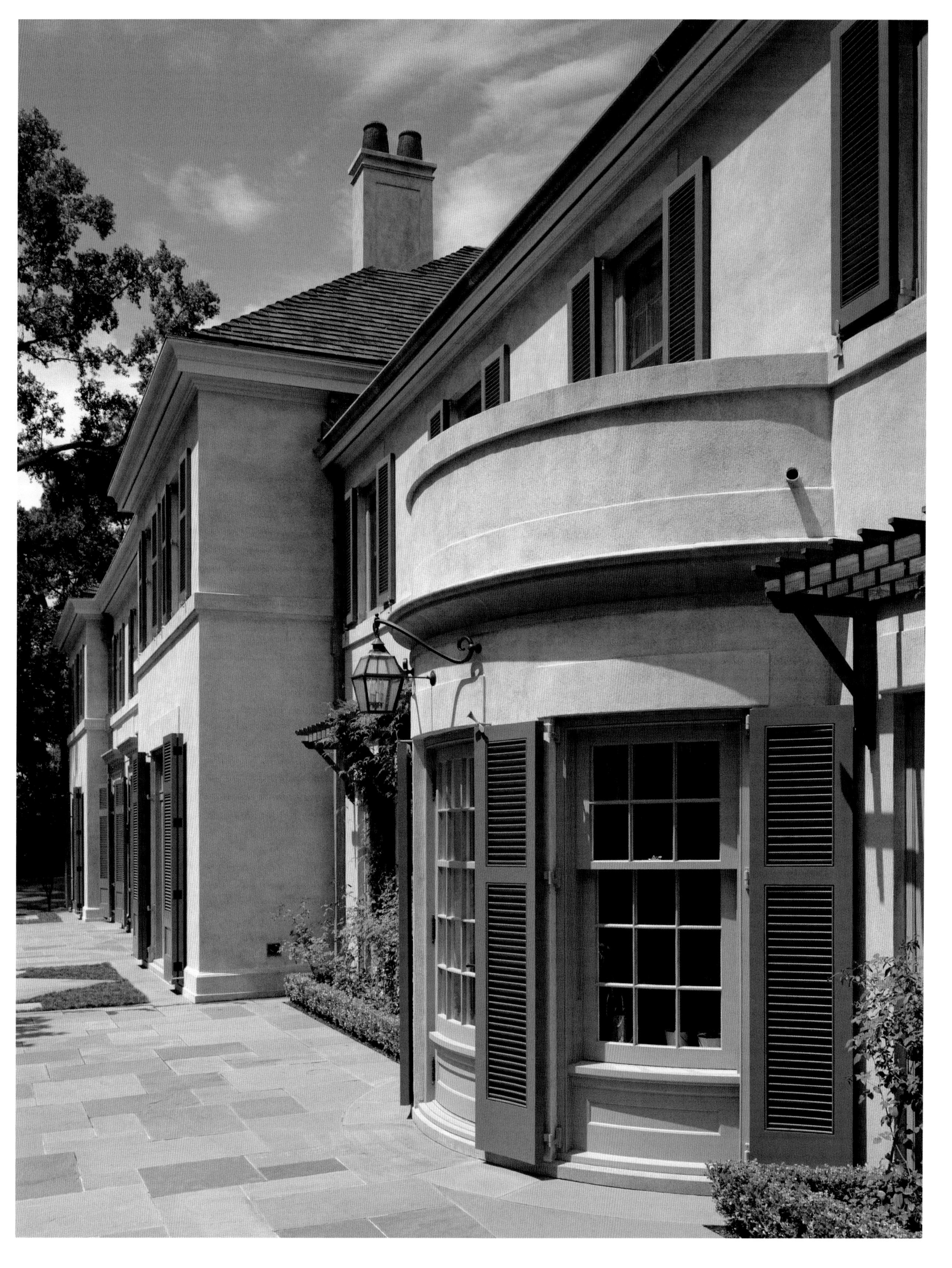

沉睡谷住宅花园
Sleepy Hollow Residence & Gardens

德克萨斯州休斯敦
Houston, Texas

该住宅始建于1929年，其选址在当时被认为是休斯敦远郊的一大块地，那是橡树河新兴花园郊区有代表性的地块。也许为了反映出这种近乎乡村的环境，建筑师弗兰克·J·福斯特设计了一套极具诺曼底地方建筑风格的法式农舍，这是他所擅长的独具特色的风格。

房屋在设计时采用了线性设计元素，有一间宽敞的屋子面向东南方，这大概是为了在没有空调的时代能够充分利用那时的季风。从最初设计这所住宅起到现在的几十年中，休斯敦城区一直在扩建，那些曾经的乡村地区现在也变成了准城市地区。

认识到这个事实，对于负责近期重新整修房屋和景观的Curtis & Windham建筑师事务所来说，设计策略中最重要的目标之一就是在房屋自身景观及其周边逐渐形成的众多建筑环境下，重新定义这座郊区住宅。

在某种程度上，用新的花园墙隔开室外区域，在房屋和景观之间设置过渡性空间，修剪现存的灌木丛底部，都属于这种策略。该项目没有破坏福斯特浓烈的个人风格和独特的法国乡土建筑特色，新的建筑和景观元素整合起来构成了从旧到新的自然过渡。

沿着前面车道新增加的砖木结构的弯曲园墙是最大亮点，它将房子定位在一个成年树冠构成的背景之中。在园墙接近中心的地方建有一个鸽房，是车道上的一个别致焦点，且和福斯特原来设计在另一处的鸽房相类似。

增加的花园墙与新的草皮网球场及汽车场一样，是为了界定各个户外花园房及其功能。建筑师在原建筑结构的主房东面增设了一间侧卧房，在南面附加了一栋L形车库、庭院和泳池。

新的庭院和泳池是连接房子和花园的带顶过渡区，这种空间类型在以前的景观设计中未曾有过。房子与景观之间的这种过渡基调也体现在其他地方，比如说那个毗邻新侧卧房的平台，就是在间隔较大的石板间铺满草皮而成；而庭院平台，也是由许多方块花圃点缀而成。

福斯特所设计的建筑反映了他对当地建材的了解和对当地手艺人才能和局限性的敏感。这栋房子是他在休斯敦仅存的一个作品，它的选材很有意思，包括炼砖、重型木和粘土屋面瓦，通过这些可以看出当地制造商提供的建材是怎样的，还有那个年代当地工匠的手艺如何。

房子和园墙所涂抹的白色涂料将它们统一成一系列互相关联的结构，同时还没有破坏那些特意砌成倾斜状的砖块的有趣纹理。

When this residence was built in 1929, this property was in what was considered to be a distant suburb of Houston and built on the type of large lot that was typical of the emerging garden suburb of River Oaks. Perhaps to reflect the almost rural setting, architect Frank J. Forster designed a French farmhouse that is reflective of Norman provincial architecture, the highly picturesque style in which he specialized. The house is a linear element in plan, one room wide and oriented to the southeast, presumably to take advantage of prevailing breezes in the days before air conditioning. Over the decades since the house was first designed, the city of Houston has expanded and the once rural setting has become quasi-urban. Identifying this fact, one of the most important objectives of the design strategy for the recent renovation of house and landscape by Curtis & Windham Architects has been to redefine this country house within its own landscape and the larger scale of neighborhood context that has evolved around it. In part, this was accomplished by defining separate outdoor zones with new garden walls, providing transitional space between house and landscape, and editing the existing understory of thickets. Working within the rich personal language established by Forster and the picturesque idiom of French Provincial architecture, new architectural and landscape elements were integrated for a seamless transition from old to new. A new curved brick and timber garden wall is the most prominent addition along the front drive that anchors the house against a backdrop of mature tree canopy. A dovecote on the approximate midway point along the wall is a picturesque focal point on the drive and corresponds to one designed by Forster in another location. Along with a new grass tennis court and motorcourt, additional garden walls were designed to define various outdoor garden rooms and functions. Additions to the main house include a bedroom wing to the east of the original structure and an L-shaped garage building, courtyard and pool to the south. The new courtyard and pool are a sheltered transition between house and garden, a type of space that was previously lacking in the landscape. This theme of transition between house and landscape is continued in the other areas; at a terrace of widely spaced flagstone interplanted with lawn adjacent to the new bedroom wing, and again at the courtyard terrace where paving is interspersed with squares filled with planting. Forster's architecture reflected thoughtfulness about local materials and sensitivity to the talents and limitations of local tradesmen. As the only surviving example of his work in Houston, it is interesting to note his selection of clinker brick, heavy timber, and clay roof tiles for this house, a reflection of the materials offered by local manufacturers and the skills of local craftsmen of that time. A coat of whitewash unifies the house and garden walls as a series of related structures without compromising the pleasing texture of the brick that is deliberately laid up out of plumb.

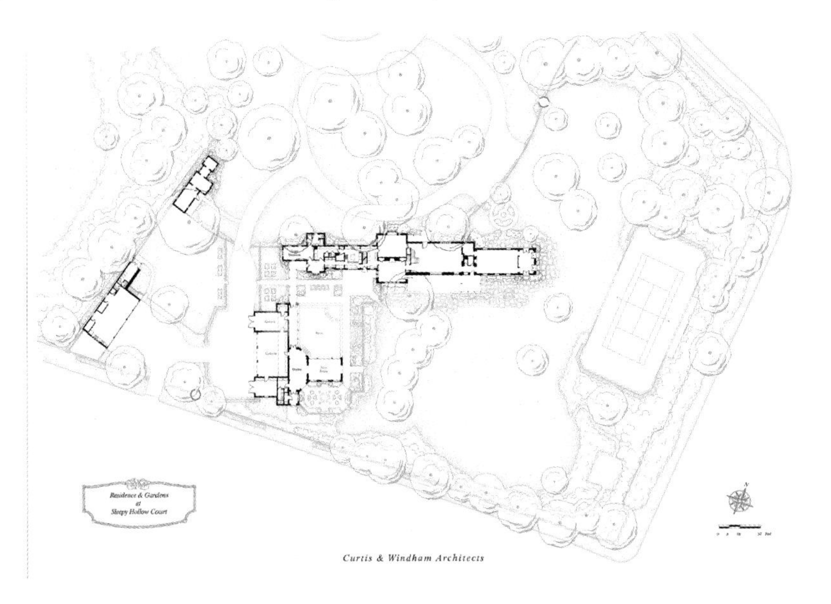

橡树河大道的花园凉亭
River Oaks Boulevard Garden Pavilion

德克萨斯州休斯敦
Houston, Texas

这座凉亭是一座原有的南方殖民地住宅的新附属建筑，该住宅位于德克萨斯州休斯敦市橡树河社区。该住宅的门廊正对着橡树河大道，该道是这个重要的休斯敦社区的第一住宅大道。通过使该凉亭的主立面与新挖的游泳池处在同一轴线上，从而将其打造成后花园的中心。建筑的辅立面（即较矮的建筑立面）上设有两扇车库门，它们被设置在一个面朝车道的设有支架的顶篷下。

花园凉亭的主立面与原有住宅相一致，都按照爱奥尼亚式风格设置了窗户。凉亭与住宅相比，维持着一个更为合适的结构高度，即在主建筑和辅建筑之间建立层级分化。在住宅中可以隐约看到花园中的建筑，尽管该建筑具有多种用途（其中设有厨房、浴室以及通往二层客房的楼梯间），但它与这里的景观却是不相宜的。一层的其他空间设有一个可停放三辆车的车库，以及一个可眺望游泳池的宽敞门廊。建筑主体按照南方殖民地建筑风格的传统，由白色的砖块砌成，并采用了木质的飞檐、支柱以及屋檐。

摄影：Hester +Hardaway Photographers

This pavilion is a new dependency built to accompany an existing Southern Colonial house in the neighborhood of River Oaks, in Houston, Texas. The portico of the main house faces River Oaks Boulevard, the premier residential boulevard of this important Houston neighborhood. The pavilion was designed to be the focal point of the back garden by orienting the primary facade on axis with the new pool. The secondary facade, the shorter elevation, has two garage doors underneath a bracketed canopy that face the drive.

The primary facade of the garden pavilion is fenestrated in the Ionic order, following the existing residence. Hierarchy is established between primary and secondary buildings by maintaining a more modest entablature height for the pavilion, in contrast to the Main House order. The garden building is viewed from the main house obliquely and serves as a folly in the landscape, while providing a multi-use program: a pool kitchen, bath and stair hall to the second floor guest suite. The remainder of the ground floor contains a three-car garage and a large porch that overlooks the pool. The body of the building is composed of white brick in the tradition of Southern Colonial architecture with the cornice, columns, and eaves made of wood.

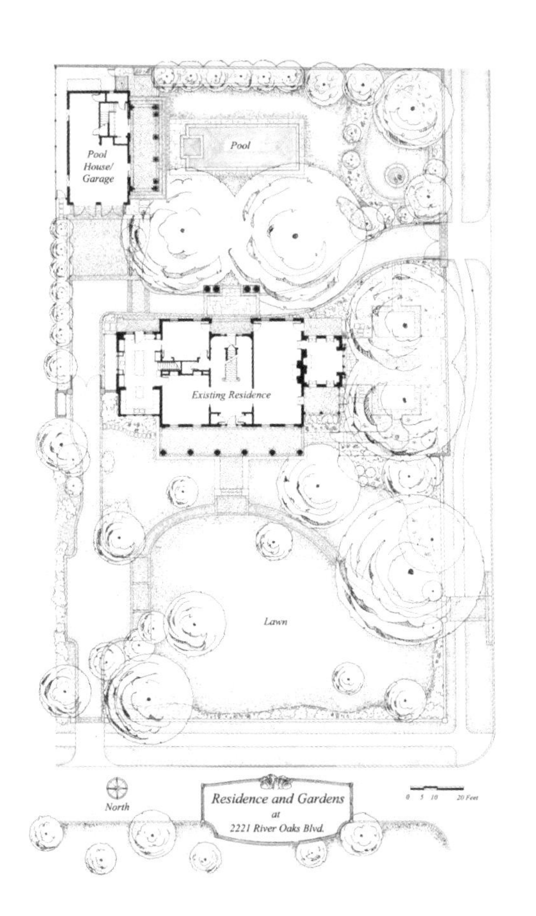
Pool
Pool House/ Garage
Existing Residence
Lawn
North
Residence and Gardens
at
2221 River Oaks Blvd.
0 5 10 20 Feet

A Garden Pavilion for Mr. and Mrs. Thomas L. Carter Jr.
Curtis & Windham
Architects

David Jones 建筑师事务所

David Jones Architects

曾几何时，你偶然瞥见一所房屋，觉得它恰到好处。房屋完善了主人的生活与品位，与场地和周围环境和谐相融，仿佛最初它就在那里。比例、细节、材料和工艺品质使得每个线条和每个转角都不可或缺。在过去与现在简单相连的共同体中，古董与现代技术交融，引人注目的艺术品与休闲花园并存。

David Jones 建筑师事务所就是用细心和创造力来打造这样的建筑。

David Jones 建筑师事务所的工作室位于华盛顿杜邦圆环社区一座联排住宅的上层，他设计的房屋遍及华盛顿及其以外的地区。

建筑师 David Jones 的背景：在普林斯顿大学获得本科及专业学位；曾在剑桥国王学院学习；曾在华盛顿的 Hartman-Cox 建筑师事务所工作。事务所的每名员工都由 David 亲自带队指导，每一位业主也会亲自参与到从最初的决定到初期和最终的设计、图纸设计、建筑商选择和建筑管理等整个流程当中。

三十多年来，该公司一直在美国享有盛誉，经常荣获设计奖项并在《传统住宅》《时期住宅》《游廊设计》和《建筑文摘》上发表文章。

Now and then you come upon a house that simply seems to fit. Complementing its owners' lives and tastes, the home occupies the land and the neighborhood comfortably, as if it has been there forever. Proportions, details, materials and quality of workmanship make every line and corner feel almost inevitable. Antiques and modern technologies, dramatic artworks and relaxing gardens coexist within an easy continuum of past and present.

With care and creativity, David Jones Architects crafts such houses.

Working in a studio atop a townhouse in Washington's Dupont Circle neighborhood, David Jones Architects designs houses that fit throughout the Washington area and beyond.

Architect David Jones's background includes undergraduate and professional degrees from Princeton; study at King's College, Cambridge; and work with Washington's Hartman-Cox Architects. Every staff of the architects works with David to guide each project, and each owner, through the processes of initial decision-making, preliminary and final design, construction drawings, builder selection and contract administration.

For over thirty years the firm has maintained a national reputation, receiving design awards with regularity along with feature articles in *Traditional Home, Period Homes, Veranda* and *Architectural Digest.*

作　品

Selected Work

滨海住宅
Beachfront House

森林山
Forest Hills

苏格兰式乔治亚庄园
Scottish Georgian Manor

圣麦克斯海滨寓所
Waterfront Retreat – St. Michaels

滨海住宅
Beachfront House

特拉华州
Delaware

这一新建的海滨度假住宅位于特拉华海岸沿线。甲方和建筑师的目的是想创造出有别于 20 世纪 40 年代被开发商弃建的那片不伦不类的小别墅群的新建筑。这种木板式建筑的灵感来自甲方在一座小镇度过的童年时光，这座小镇因 H.H.Richardson 的一系列建筑而闻名。

住宅建于街角，海滩和街道的景观一览无遗。建筑师利用这种双重性设计了两种不同的立面——一面是宽阔的山形墙，深邃的长廊延伸至东侧的沙滩和大海，而另一个则是更加垂直的塔形 / 山形墙复合结构，引人至街道和北侧的海滩小径。

整齐的板条和多样化的层次使木板立面更显生动。多窗格的窗户锁定了风景并增添了围合的感觉。从入口楼梯处以灯塔为灵感的螺旋楼梯中柱开始，航海主题既运用到嵌板的装饰上，同时也贯穿所有的细节。

This new oceanfront vacation home is located along a popular stretch of the Delaware coast. The goal of the owner and architect was to create an architectural statement amongst the nondescript cottages thrown up by developers since the 1940s. The inspiration for this shingle style residence was derived from the owner's own childhood in a town renowned for its collection of H.H. Richardson buildings.

Situated on a corner lot, the house presides over both the beach and the street. The architects used this duality to create two different facades—a broad gable with a deep porch addresses the beach and ocean to the east, while a more vertical tower/gable composition addresses the street and beach path on the north side.

Trim bands and variations in coursing enliven the shingled facades. Multi-paned windows frame views and give a sense of enclosure. Nautical motifs decorate panels and inform the details, beginning with the lighthouse-inspired newel post of the entry stair.

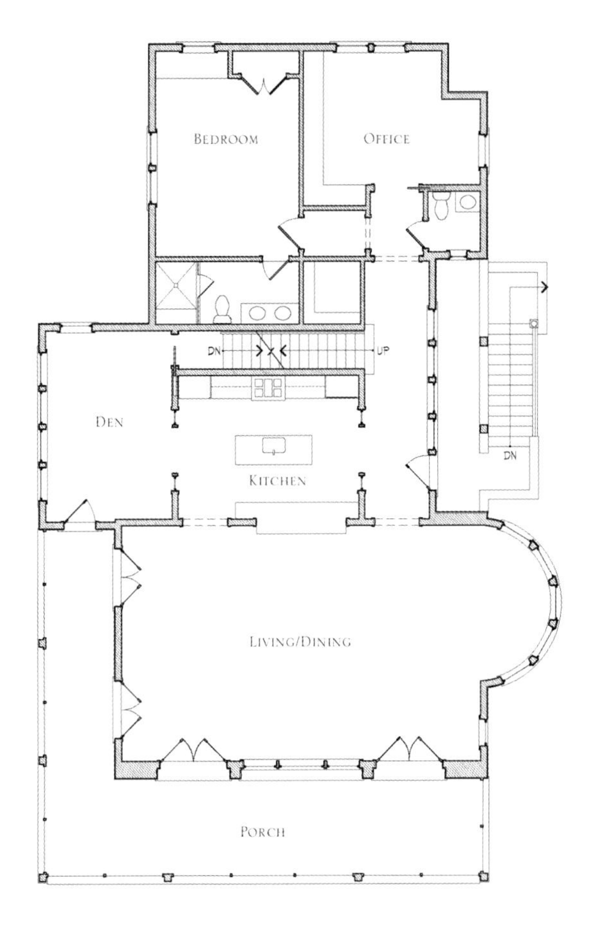

First Floor Plan

森林山
Forest Hills

华盛顿
Washington

这个家庭需要新建一所住宅，能够满足偶尔招待客人的需求并适应孩子活泼好动的个性。父母则各自想要一个能够逃离现实的私人空间——妈妈需要一间书房，爸爸要一间藏书室。他们想要的住宅要有足够的户外空间，可以修建一块平坦的草坪用来做运动，私人车道上建一处半场篮球场，此外还有一汪池塘。

他们在华盛顿某建成小区中发现了一块极具价值的空地，可俯视连绵起伏的峡谷。这块地从一处较大的房产中分割出来，该房产曾卷入20世纪20年代的大开发中。具有历史意义的19世纪著名木板式房屋"猫头鹰之巢"就位于场地的旁边。因此，这一特别的街区在严格的历史保护条例的保护下留存至今。

这一新建住宅延续了"猫头鹰之巢"的风格，但却避免复制其特征和形式。与旁边具有历史意义的"猫头鹰之巢"一样，它也采用木板式结构和石材基础。尽管住宅的体量较大，但是它被分解为目前的小规模空间，掩饰了其实际的建筑面积。

在场地的前面，住宅体量构成了浪漫的街道景观：引人注目的山形墙遮掩了凹陷的入口走廊，角楼和低矮的车库翼楼一直延伸到东侧。在朝向"猫头鹰之巢"的场地西侧，屋顶向下倾斜，将藏书室覆盖，以与其具有历史意义的邻居相呼应。

在场地的后方，成对的山形墙容纳三层的客房，并将视线引至草坪和远处的峡谷。宽敞的厨房和家庭娱乐室朝后部的长廊和露台开放，透过窗户和开间可以看到街道景观。

内部，两个楼梯间（前厅与后厅）将一层与楼上的主人卧室和地下室的娱乐室相连。藏书室远离起居室，而书房建于车库之上，甚为幽静，为丈夫和妻子分别提供了庇护之所。从每个房间都能看见前后院，因此父母能够掌握孩子们的活动情况。

This family needed a new house to accommodate casual entertaining and their active children. The parents each asked for a private get-away – a study for her and a library for him. They sought a property with enough outdoor space for a level lawn for sports, a ½ basketball court in the driveway, and a pool.

They found a prized vacant lot overlooking a rolling valley in an established Washington DC neighborhood. The lot was newly carved from a large estate that had been engulfed by development in the 1920s. The "Owl's Nest", an historic 19th Century shingle style mansion, looms next door. As a result, this particular block comes under strict historic preservation covenants.

This new house takes its stylistic cues from the Owl's Nest, but replication of features and forms is avoided. Like its historic neighbor, this new house is rendered in the shingle style with a stone base. Although it is large, the massing of the house is broken down to present a smaller scale which belies the actual square footage.

On the front, volumes are arranged in a romantic streetscape: a prominent gable sheltering the recessed entry porch, a turret, and a low garage wing extending to the east. Towards the Owl's Nest on the west, the roof swoops down over the library in deference to the historic neighbor.

At the rear, paired gables accommodate third floor guest rooms and announce the view to the lawn and the valley beyond. A commodious kitchen and family room open onto the rear porch and terrace and are visually connected to the landscape with a bank of windows and a large bay.

Inside, two staircases (front hall & back hall) connect the first floor with the family bedrooms above and the basement recreation room. The library off the living room and the remote study above the garage offer refuge for the husband and wife, respectively. Each room offers views to the front and rear so parents can keep abreast of activities.

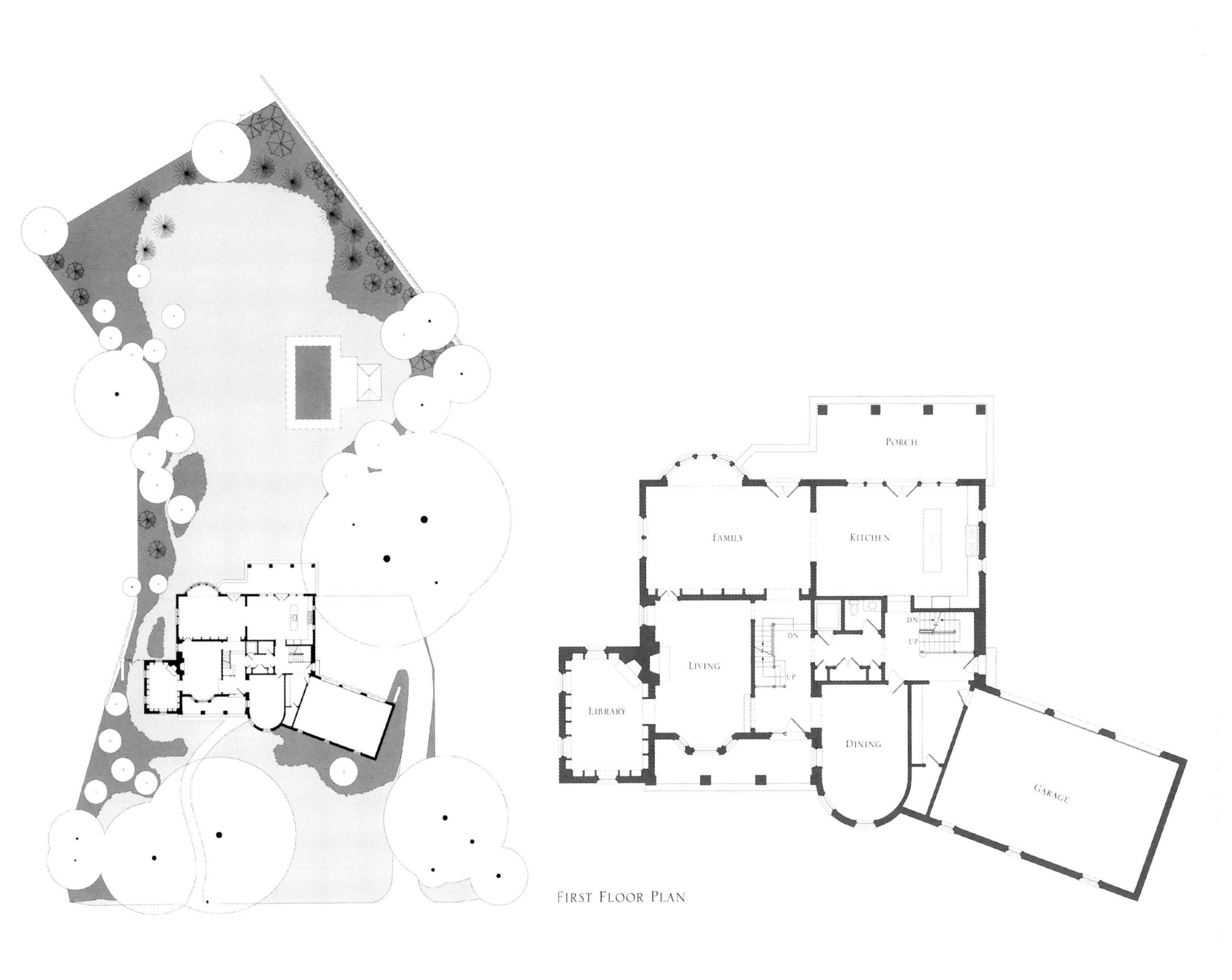

First Floor Plan

苏格兰式乔治亚庄园
Scottish Georgian Manor

纽约市
New York City

这是一所能够满足经常性的客人来访和频繁的娱乐聚会要求的住宅，带有四车位的车库、一间娱乐媒体室、远离主人三间卧室的若干客房。住宅应明亮、宽敞，方便进入屋后的露台和草坪。

建筑师和甲方对延续美国传统的住宅设计有着共同的兴趣。他们尤其痴迷于费城外斯古吉尔河岸的“欢乐山”乡村住宅，这所住宅由一位苏格兰船长在1761年建造。

建筑师的策略是利用分为五部分的乔治亚式综合体来满足现代家庭的日常功能，房屋中央双层主体的灵感则来自“欢乐山”。中央区域的进深也增加了——采用双房间进深取代原有的单房间进深——以容纳更多的功能空间。从前到后的轴远景被保留了下来。横向的楼梯间将翼楼连接起来，同时可使阳光通过窗户照入房屋中央。

侧面带拱廊的翼楼带有封闭的外立面，掩饰了后方房间的实际进深。透过宽阔的窗户俯视屋后的露台和花园是一侧的大型娱乐室和另一侧家庭厨房的特色。两个对称端的阁楼与建筑主体相呼应，并巧妙处理了楼层间的高差。每座阁楼的一层都建有地上车库，车库上层是客房和家庭办公室。

摄影：Seong Kwon

A large house to accommodate extended family visits and frequent entertaining, with garage space for 4 cars, a party media room, and guest bedrooms located away from the family's three bedrooms. The house should be light-filled and spacious with easy access to a rear terrace and lawn.

The architect and clients share an interest in the continuation of traditional residential design in America. In particular, they are drawn to "Mount Pleasant," a Georgian country residence on the Schuylkill River outside of Philadelphia, built by a Scottish sea captain in 1761.

The architect's strategy is to accommodate the family's modern day program within a Georgian five-part composition, with Mount Pleasant serving as the inspiration for the central two-story body of the house.The depth of this central block is increased—two rooms deep instead of the original's one—to house the extensive program. The axial vista from front to back is maintained, while a new transverse stair hall provides connection to the wings and allows windows to flood light to the center of the house.

Flanking arcaded wings with closed formal facades mask the true depth of the rooms behind. A large party room on one side and the family kitchen on the other feature large expanses of windows overlooking the terraces and garden to the rear. The symmetrical end pavilions echo the main block, yet manipulate floor levels. Each pavilion contains garages on the ground level, with guest rooms and a home office above.

1. ENTRY HALL 2. DINING ROOM 3. LIBRARY 4. LIVING ROOM 5. FAMILY ROOM 6. KITCHEN 7. PARTY ROOM 8. GARAGE

1. Entry Hall 2. Dining Room 3. Library 4. Living Room 5. Family Room 6. Kitchen 7. Party Room 8. Garage 9. Master Bedroom 10. Master Sitting Room 11. Master Bath 12. Bedroom 13. Office 14. Pool House

圣麦克斯海滨寓所
Waterfront Retreat - St. Michaels

组约市

New York City

甲方被海滨田园环境所吸引，于 2005 年购买了此处房产。场地位于道路的末端，是流入切萨皮克湾一支支流的转弯处，可将落日余晖尽洒水面的景色与如画的水路远景尽收眼底。遗憾的是，与房产同时购得的住宅却不尽如人意。这座自 20 世纪 50 年代就存在的住宅好是好，但却不激动人心。侧面的车库和日光室以及屋后的开间，作为房屋的附属十分不雅观。天花板低矮，内部房间拥挤杂乱。建筑师的任务就是将原有的住宅尽可能地保留下来，并将在东海岸社区具有历史价值的“丑小鸭”变成优雅的“白天鹅”。

甲方需要较为开放的楼层设计，尽可能地增加层高，并在二层设置主卧，以欣赏水上景色。为了达到这一目的，建筑师围绕着重新设计的中心安排了楼层平面和体量。在平面中，建筑师开辟出一间大型的对称式起居室 / 餐厅，并新增一条可以俯瞰水景的走廊。在建筑体量方面，住宅主体的屋顶被提高以增加一层和二层新建主卧的天花板高度。原有走廊新增的屋顶采光窗和栏杆凸显了住宅前方的景色。主卧凸起的烟囱和带有人字墙的走廊与长长的滨水立面形成了紧密的联系。

为了使整个建筑和谐统一，砖墙、侧面护墙板、木质装饰一律粉刷成白色。新增的百叶窗和木板屋顶使建筑具有独特的质地与个性。新的池塘以及池塘凉亭建在家庭娱乐室附近，可将水景尽收眼底。新建的护墙板附属建筑将池塘区域与车库和车道隔离开来。即便如此，住宅的改建仍基于之前的先例。新增的地热泵用于供暖和制冷——是对高利用效率的现代生活的反映。

摄影：Seong Kwon

The clients purchased this property in 2005, charmed by the idyllic waterfront setting. Situated at the end of the road and at the bend in a tributary feeding into Chesapeake Bay, it offers a view of the sunset across the water and picturesque vistas down the waterway. Regrettably, the house that came with the property was not so charming. The 1950's house was sound but uninspiring. Unsightly additions had been attached—a carport to the side and sunrooms and a bay along the back. The ceilings were low, and the interior was a warren of rooms. The architect's assignment was to keep as much of the existing house as possible while turning this ugly duckling into a swan in this historic Eastern Shore community.

The clients asked for a more open plan, higher ceilings where possible and a 2nd floor master bedroom with a view of the water. To achieve this, the architect organized the plan and massing around a re-created center. In plan, a large symmetrical living/dining room is carved out and a new porch overlooking the water is added. In massing, the roof of the center block is raised to provide for a higher ceiling at the first floor and the new master bedroom above. New dormers and a new balustrade over the existing porch give focus to the front, while raised chimneys and a pedimented porch at the master bedroom create a dramatic anchor to the long waterfront elevation.

To unify the composition, the brick walls, clapboard siding and wood trim are all painted white. Shutters are added and a new wood shingle roof lends texture and character. The new pool and pool pavilion are placed to the side near the family room, allowing for uninterrupted water views from the house. New clapboard outbuildings shield the pool area from the garage and driveway. Even though this remodeling is based on traditional precedent, new geothermal heat pumps provide heating and cooling—a reflection of present day high utility costs.

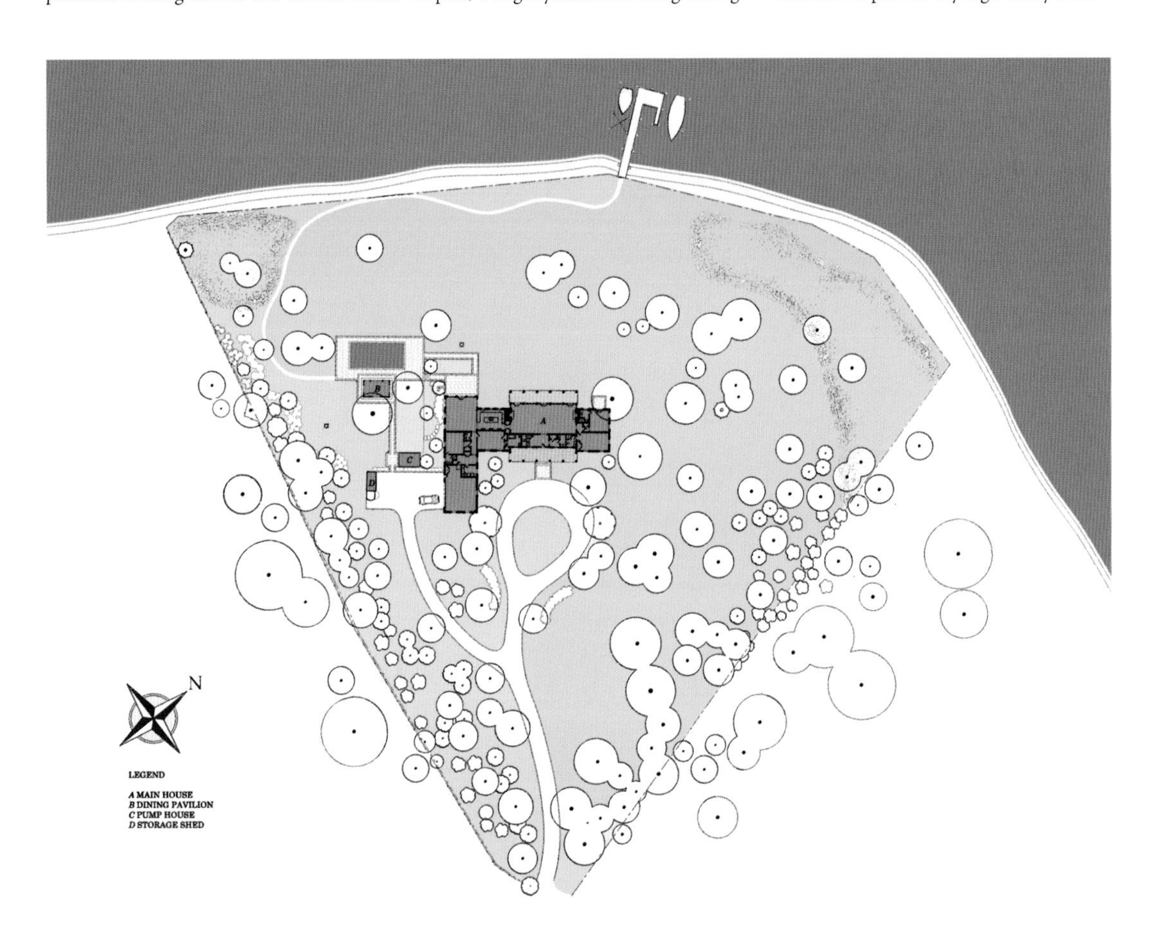

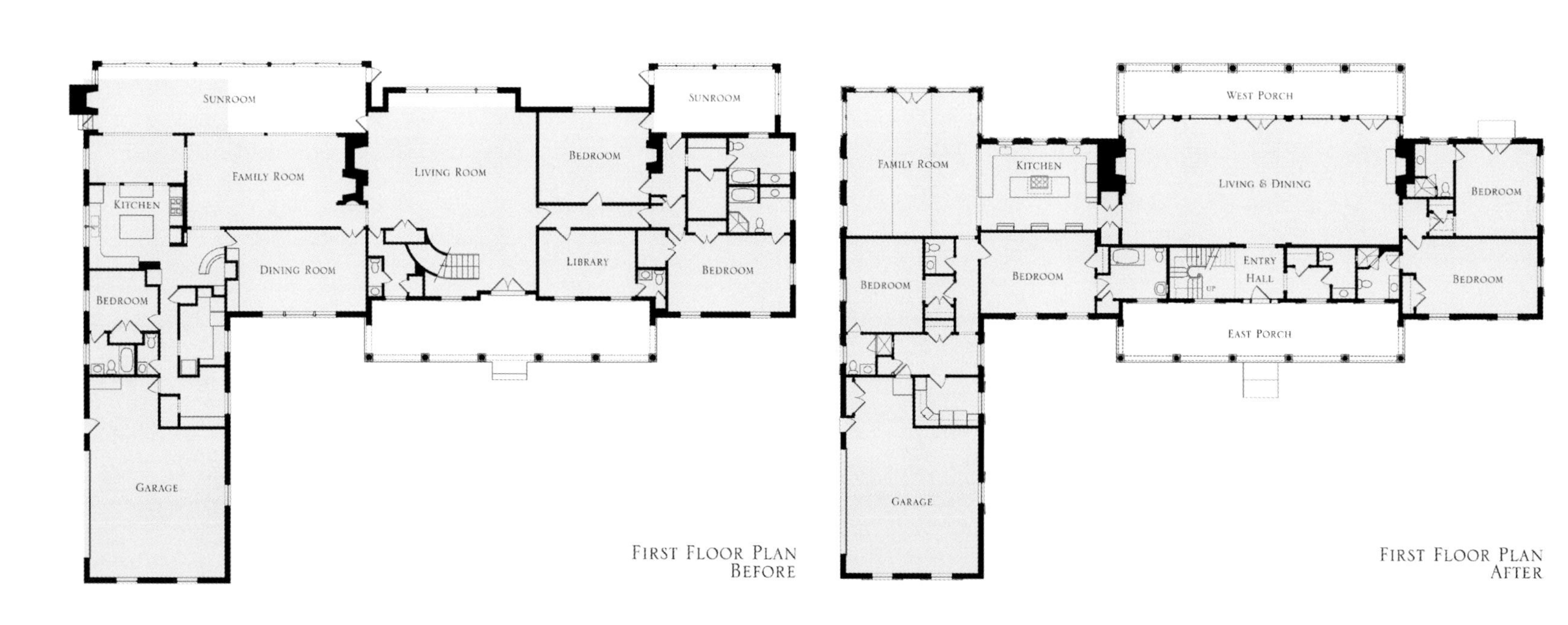

First Floor Plan
Before

First Floor Plan
After

Derrick 建筑事务所

Derrick Architecture

Christopher Derrick 是 Derrick 建筑事务所的创始人和首席建筑师，于芝加哥伊利诺斯州州立大学获得建筑结构学学士学位（五年制）。在大学期间，Christopher 是荣誉学院的成员，同时也是 John E. Walley 奖学金的获得者。在开创自己的公司以前，他在芝加哥和佛罗里达地区的商业和住宅建筑公司工作。Christopher 开创的 Derrick 建筑事务所专门从事住宅设计、翻修与室内设计工程。Derrick 建筑事务所的首要目标是通过高标准的设计品质建造贴近甲方品位和预算的住宅，为甲方提供周到、专注的专业服务。

自大学时期起，Christopher 对传统和古典建筑的热爱就一直激励他个人不断地对这一学科进行学习和研究。由于他对建筑环境的关注，他曾经就任第一社区的副主席，第一社区是一个鼓励对已建社区进行兼容再开发的组织。他是古典建筑与艺术学院芝加哥中西部分会的创始董事。

Christopher Derrick – Principal and founder of Derrick Architecture, received his Bachelor of Architecture in Structures (5-Year) from the University of Illinois at Chicago. While attending UIC, Christopher was a member of the Honors College and recipient of the John E. Walley Memorial Scholarship. Before starting his own firm, Christopher worked in both Commercial and Residential Architecture firms in the Chicago and Florida areas. Christopher began Derrick Architecture as an architectural firm specializing in residential design, renovation and interiors. Derrick Architecture's primary goal is to provide thoughtful and dedicated professional service to its clients by producing the highest levels of design quality possible while adhering to its clients' tastes and budget.

Since college, his love for Traditional and Classical architecture has prompted him to continuing personal research and study of the subject. Because of his concern for the architectural environment, he has served in the past as Vice President of Community First, an organization encouraging compatible redevelopment of established communities. He is also the current and founding President of the Chicago-Midwest Chapter of the Institute of Classical Architecture and Art.

作 品

Selected Work

绿色田野住宅

Green Acres House

伊利诺斯州内伯威尔市

Naperville, Illinois

绿色田野是为家庭的享受和便利而设计的住宅。位于中心的旋转楼梯既是平面布置的重点又是住宅设计的一个基础建筑元素。所有房间都由中央楼梯间向四周辐射，形成了二层的走廊。这条走廊仿佛不是建造而成的，而是雕刻出来的。每个房间的布局都考虑到营造出一种穿越住宅的连续性体验。托斯卡纳圆柱整齐地立于拱形入口两侧，创造出从一个房间到另一个房间的视觉和真实大门的效果。起居空间位于轴线之上，其主要入口在楼梯对面。起居空间是家庭活动的主要场所。住宅同时设有多样的户外空间，如前方可供用餐的走廊和后面的露台，都是起居空间的扩展。

宽敞的卧室为每一位家庭成员提供了足够的休息之所。每间卧室设有单独的卫浴。

住宅外部采用砖石构造，木材设计贯穿所有的细部。设计师精心打造了每个建筑立面，为客人同样也为家庭成员营造视觉吸引力。多种风格的建筑元素紧密结合，成就了该建筑完美永恒的设计。

摄影：Bill Meyer Photography

Green Acres is a house designed for the luxury and convenience of a family. The centrally located circular stair serves as a focal point for the floor plans and a foundational architectural element for the home. All the rooms radiate off this central stair hall, creating a second floor hallway that looks carved more than built. Each room is located with views in mind to create a sequential experience as one walks through this residence. Arched entries flanked by Tuscan columns align to create visual as well as actual portals from room to room. The great room is located on axis with the main entry on the opposite side of the stairs, acting as an anchor to ground the family's activities.

Various outdoor spaces, such as the dining porch in the front, or the veranda at the rear, become extensions of the living space.

Spacious bed rooms provide plenty of room for each member of this family. Each bedroom has its own private bath.

The exterior of this home is comprised of stone and brick, with wood architectural accent details throughout. Each elevation is carefully crafted to create visual interest for guests and family members alike. Architectural elements have been borrowed from different styles to create this cohesive and timeless design.

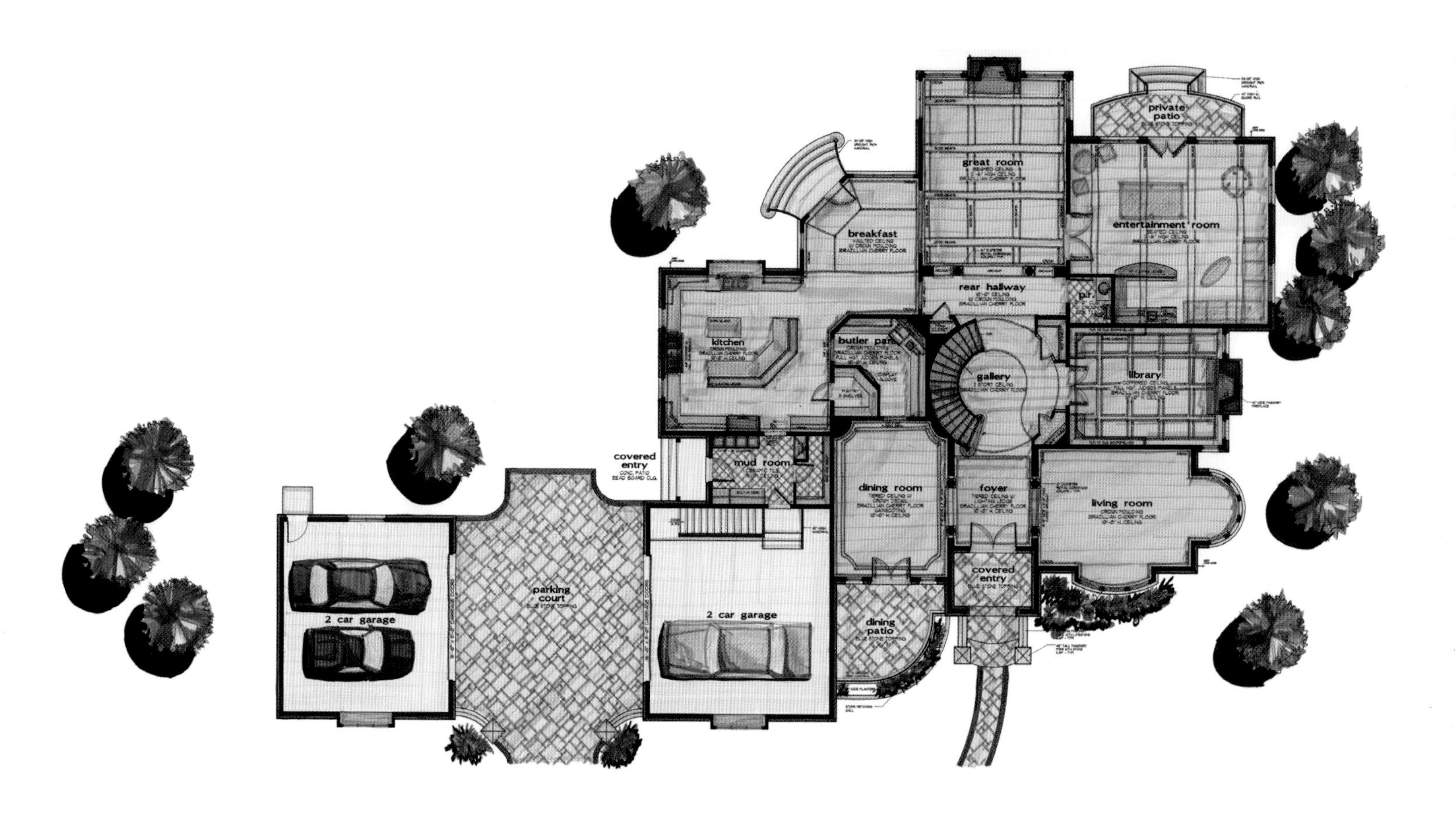

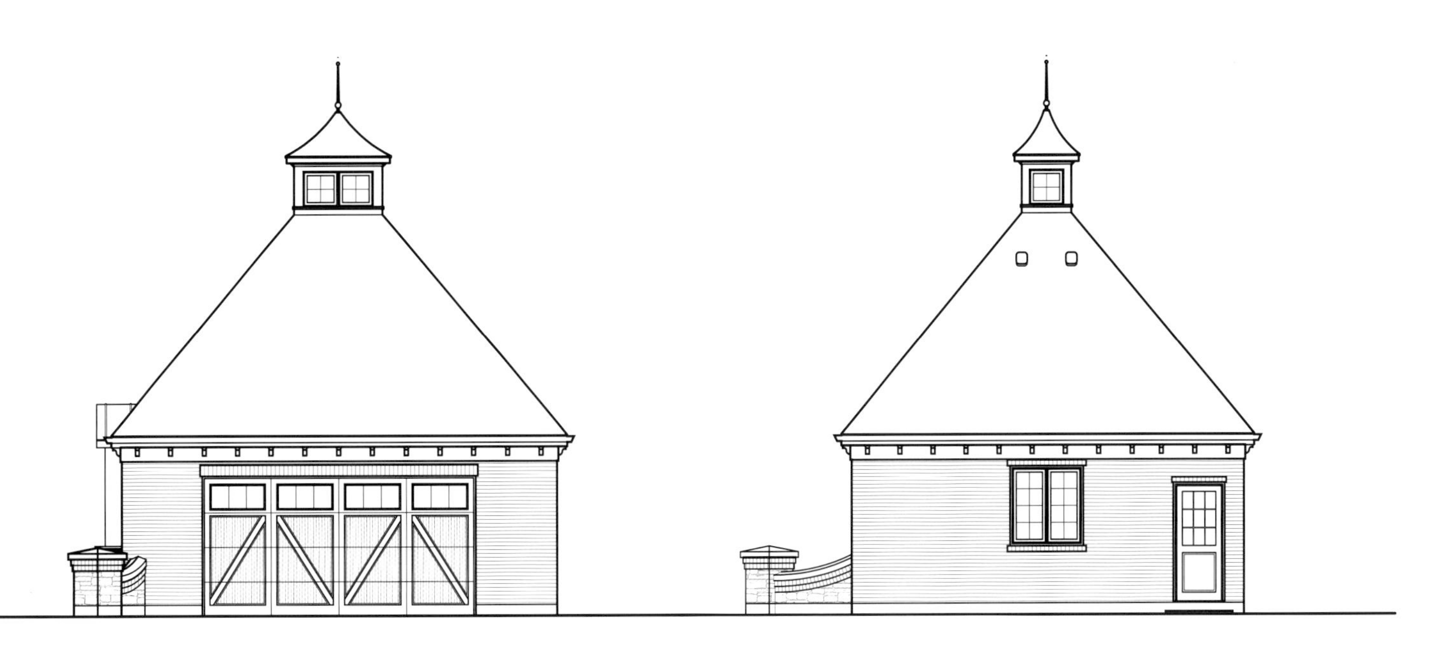

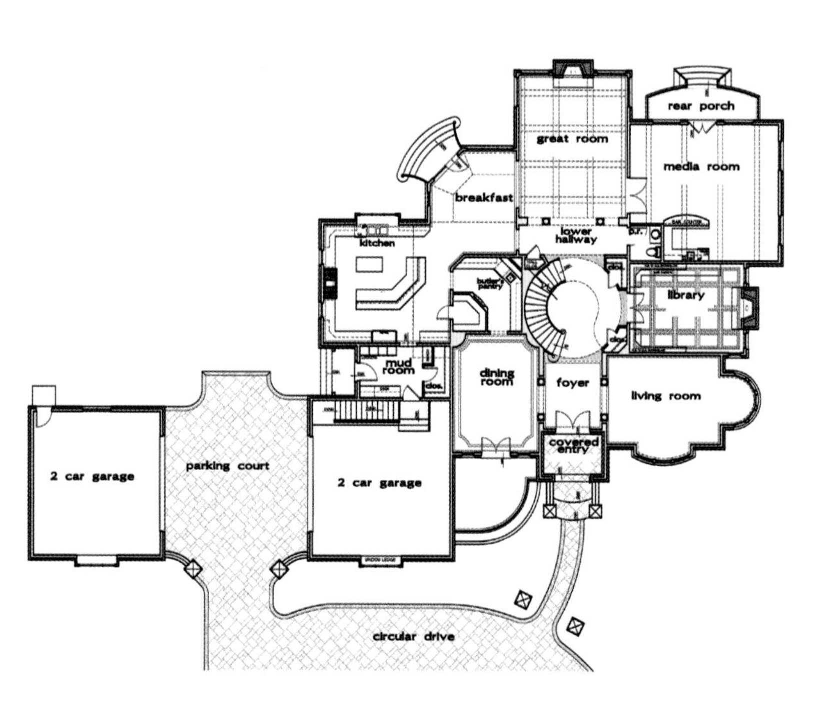

rear porch
great room
media room
breakfast
kitchen
lower hallway
library
mud room
dining room
foyer
living room
covered entry
2 car garage
parking court
2 car garage
circular drive

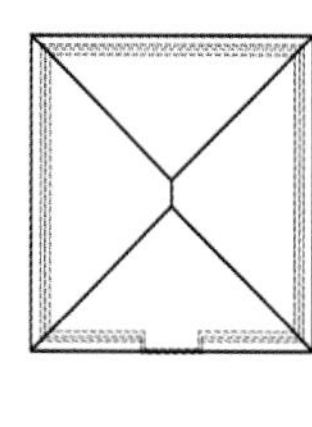

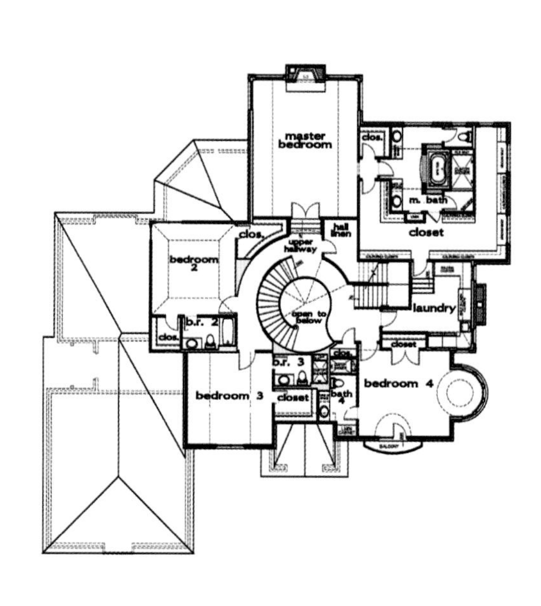

master bedroom
m. bath
closet
bedroom 2
upper hallway
laundry
b.r. 2
open to below
bedroom 3
b.r. 3
closet
bedroom 4

TUTANKHAMUN

高地住宅
Highlands House

伊利诺斯州内伯威尔市
Naperville, Illinois

这所住宅是对木板式住宅的现代诠释。甲方的意图是使之与周边现有的住宅能够和谐统一。高地住宅采用丰迪拉克礁石作为主要楼层的基础，之上的楼层采用搭叠的雪松侧面板作为过渡。铜质直立缝细节标记出楼层间的过渡。主要的山形墙入口建有突出的铜质屋顶，将客人引至木兰树阴遮蔽下的前门廊。前方入口的双重山形墙延长了建筑，因而整个住宅更加显眼，并将重点放在了水平方向的设计上。

进入住宅，首先通过的空间是楼梯间，平缓的楼梯将人引至上层。正式的餐厅贴以带图案的壁纸，人工修饰的天花板饰以简单的边线和花纹细节。餐具室连接着餐厅和厨房，同时可以作为小酒吧以供娱乐。厨房门面向起居室，起居室又可通往招待客人的集会厅或音乐厅。整面墙板以壁柱作为边框，使天花板檐口别具情调，为乐迷创造了理想环境。

楼层高度很高的日光浴室向带顶的走廊开放，营造向室外开敞的感觉。需要较大空间时，可以打开小门将日光浴室与起居室连接起来。在起居室中，天花板的风化横梁围合着装饰性的天花板镶板，大壁炉置于房中央。藏书室提供了躲避日间劳碌之所，视线及至修葺整齐的花园。

住宅的二层和三层是主人套房和另外三间带有单独卫浴的卧室。

高地住宅弯曲的楼面设计是该住宅典型的建筑风格。除了风格一致的装饰物，所有的角落、开间、屋顶采光窗都赋予这所精美住宅独一无二的品质。

This modern interpretation of the shingle style house began with the homeowner's intent of a home that was compatible with the existing neighborhood. The Highlands has a Fond du Lac ledge stone base at the main floor, transitioning to lapped cedar shingle siding for the floors above. A standing seam copper detail marks the transition between the floors. The main gabled entry with a copper roof projection invites guests onto the front porch in the shadow of the magnolia trees. The double gables at the front entry continue the length of the structure, thus emanating a pronounced mass and a major emphasis on horizontality.

Entering the house, one is led to the stair hall with a gently swooping stair inviting you to the floors above. The formal dining room reveals patterned wall paper and a faux finished ceiling framed by a simple tier and crown detail. The butler pantry connecting the dining room and kitchen doubles as a bar for entertaining. The kitchen opens to the great hall which leads to the gathering hall, or music room, where guests are received. Full height wall paneling is framed by pilasters engaging the ceiling cornice creating an ideal setting for the musical aficionado.

The ceiling soars at the solarium creating an outdoor feeling which opens onto the covered porch. Pocket doors expand the solarium into the great room when larger spaces become necessary. In the Great room, weathered beams in the ceiling frame the ornamental ceiling panels centering the main fireplace. The library provides a place of refuge from the business of the day, with views looking out over the finely manicured garden.

The second and third floors house the Master suite and an additional 3 bedrooms with private baths.

The meandering floor plan of the Highlands is typical for this house's style. All the nooks, bays and dormers in addition to the consistent ornamentation give this fine house a character that cannot be duplicated.

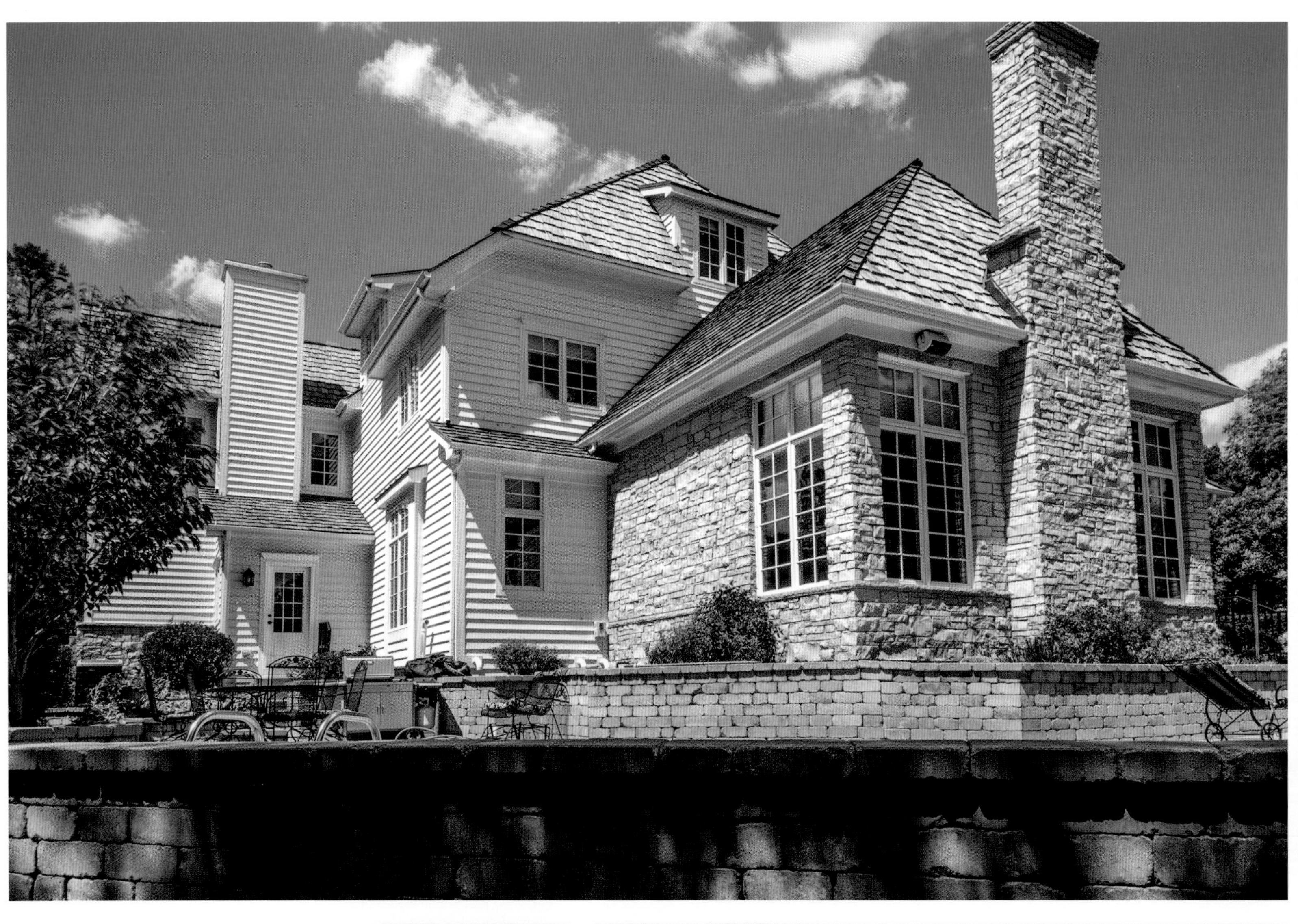

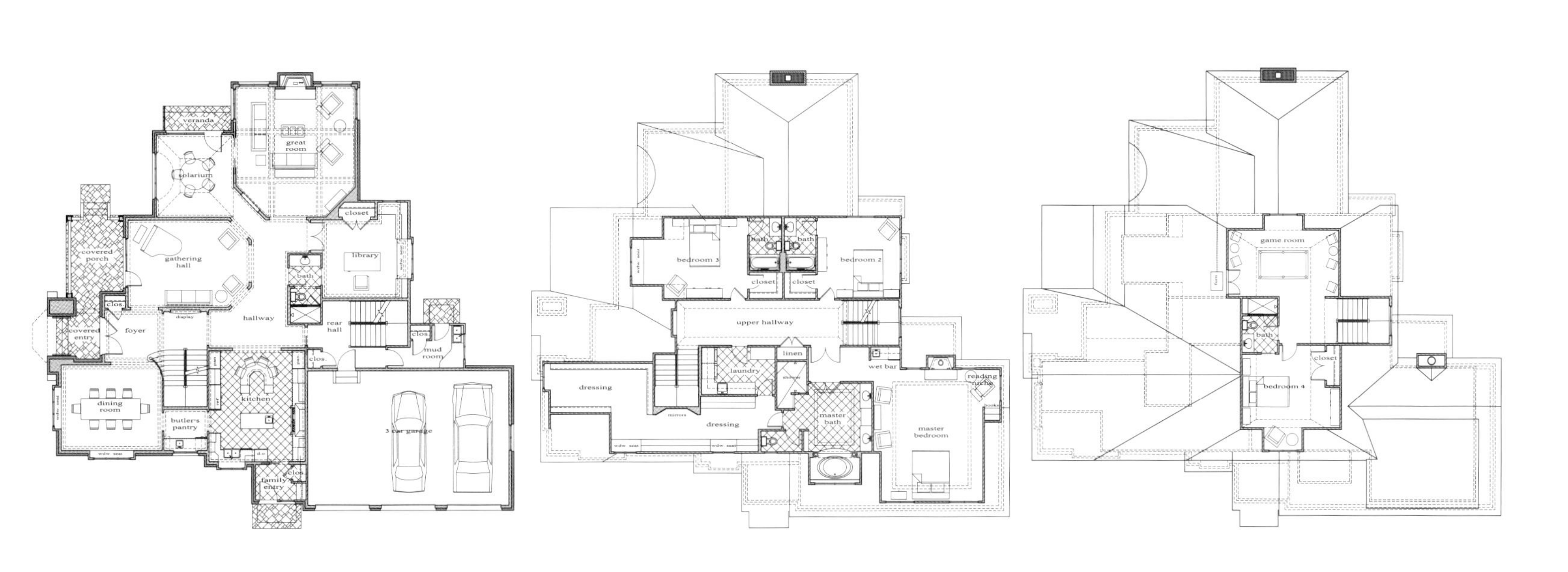
veranda
great room
solarium
closet
covered porch
gathering hall
library
bath
clos.
display
hallway
rear hall
covered entry
foyer
clos
mud room
dining room
wdw. seat
butler's pantry
kitchen
3 car garage
family entry
bedroom 3
bath
bath
bedroom 2
closet
closet
upper hallway
linen
laundry
wet bar
reading niche
dressing
mirrors
dressing
wdw. seat
wdw. seat
master bath
master bedroom
game room
bath
closet
bedroom 4

欧扎克湾住宅
Ozark Cove House

密苏里州欧扎克湖
Lake Ozark, Missouri

建筑师在设计欧扎克湾住宅时将隐私和对美的追求铭记于心。住宅围绕欧扎克湖的私人河湾建造，位于面积约为 20 234m^2 的土地上，可俯瞰水景。设计要求房间环抱自然地势，尽可能从每个房间都可欣赏到水景。住宅后面由走廊、阳台、凉廊围绕，以欣赏全年不同季节的风光。住宅位于水位线约12m之上，走出后院就能通过缓坡步入欧扎克湖。位置较低的池塘露台不会阻挡视野，不显眼的边缘也能突出池塘前的开阔视野。一处集水区聚集池塘中的水，形成了一挂瀑布。在池塘的一端建有下沉式厨房，其上是用裸露的雪松做椽尾的小屋，厨房同时可作池滨酒吧使用。池塘另一边设有热水浴桶，俯瞰池塘和水湾。五间卧室除设有私人卫浴外，还单独设有阳台或者露台。

开放式的楼面设计使人联想到住宅所处的户外环境。透过宽阔的大窗，可将住宅内外的自然之美尽收眼底。装潢奢侈的酒吧在主楼层和较低的楼层，与池塘酒吧一起，使整个住宅成为理想的娱乐之所。石材和雪松木板的选择增添了住宅的天然感觉，似乎成为自然景观的一部分。天然的雪松色调和修剪整齐的深绿色植物更强化了这一效果。

住宅深受其环境影响，并与之融为一体。它是隐匿于美景之中的居所，其设计风格与大自然不可分割。住宅本身给人以平静、放松之感。

Ozark Cove is a home design with privacy and beauty in mind. This house overlooks the water on a 20,234m^2 parcel surrounding a private cove in the Lake of the Ozarks. The design calls for rooms that hug the natural terrain, allowing views across the lake from nearly every room. The rear of the home is wrapped with porches, balconies and verandas to enjoy the scenery year round. Sitting 12m above the water line, the house gradually terraces down to the water as soon as you walk out the rear. A depressed pool patio allows unobstructed views, while an invisible edge creates the illusion of a continuous horizon from the pool. A watershed catches the water from the pool as it turns into a waterfall. At the pool's one end, a sunken kitchen sits under a cabana with exposed cedar rafter tails and doubles as a swim up bar for the pool, while at the other end a hot tub is perched to overlook the pool and cove. Each of the 5 bedrooms has a private balcony or terrace in addition to a private bath. The open floor plan is reminiscent of the outdoors this house is settled in. Large expanses of windows capture the beauty of nature, both inside and out. Extravagant bars on both the main and lower floors, as well as the pool bar, make this house ideal for entertaining. The choice of stone and cedar shakes has a natural feel which makes the home appear to be a part of the landscape. The natural cedar hues and the deep green trim further this effect. This home is greatly influenced by its context. It seems to belong in its pristine setting. Not only is this home a refuge nestled in the landscape, but its design contends to be part of the nature. The house itself seems to have taken on a peaceful and relaxing air.

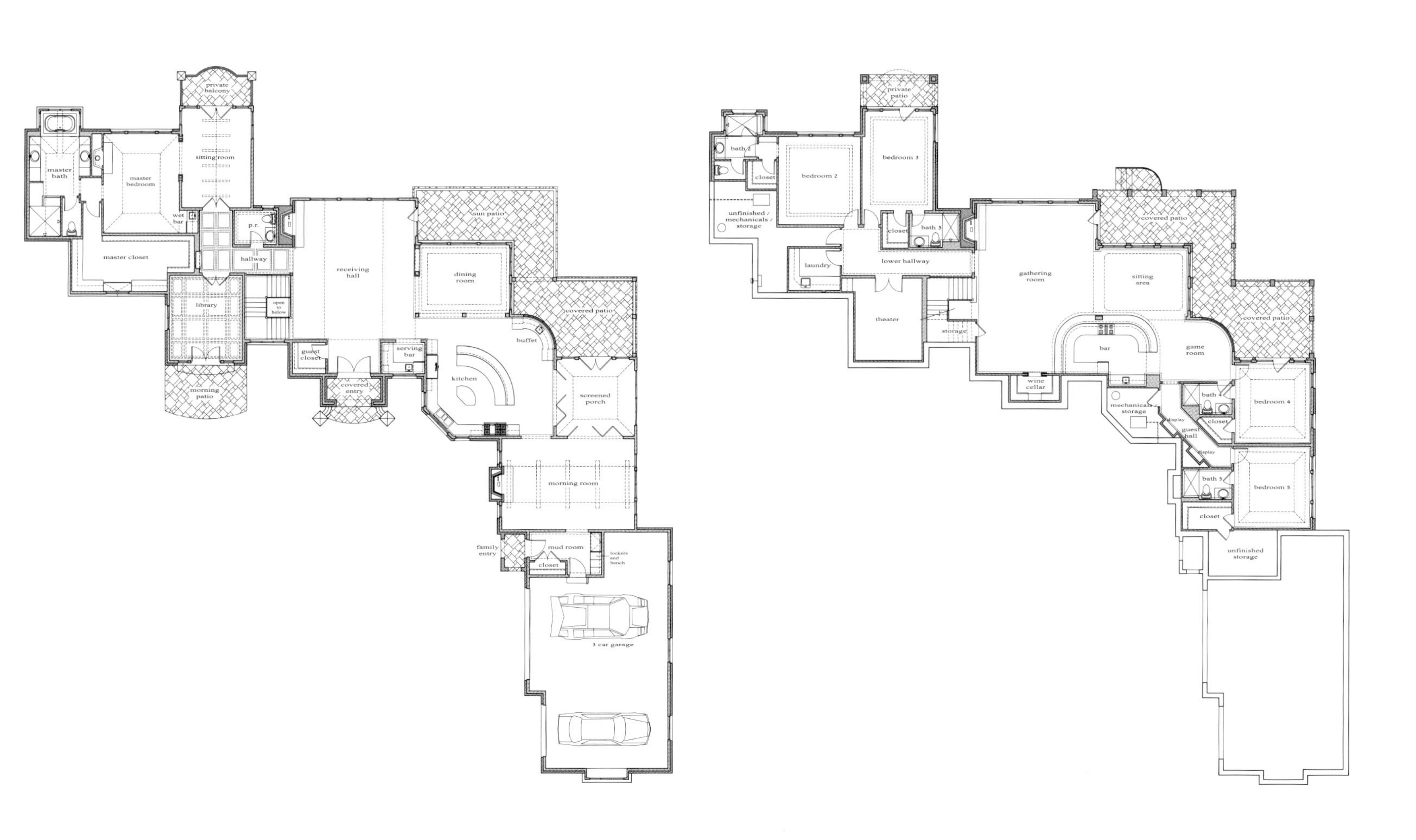
private balcony
sitting room
master bath
master bedroom
wet bar
p.r.
hallway
master closet
library
open to below
receiving hall
sun patio
dining room
covered patio
buffet
guest closet
serving bar
kitchen
screened porch
covered entry
morning patio
morning room
family entry
mud room
lockers and bench
closet
3 car garage
private patio
bath 2
closet
bedroom 2
bedroom 3
unfinished / mechanicals / storage
closet
bath 3
covered patio
laundry
lower hallway
gathering room
sitting area
theater
storage
covered patio
bar
game room
wine cellar
mechanicals storage
bath 4
bedroom 4
closet
guest hall
display
bath 5
bedroom 5
closet
unfinished storage

落日住宅
Sunset House

伊利诺斯州 LaGrange
LaGrange, Illinois

该住宅的设计灵感来自法国文艺复兴时期传统建筑的永恒吸引力，带石灰岩雕刻的砖墙庄严肃穆，着力传达出其持久性与重要性。石灰石带状层将住宅的一层和二层分隔开来，前门入口处的细节突出显示出其地位的重要性。对称的外立面有着陡峭的斜屋顶和房檐的喇叭造型，在人接近住宅时营造出令人愉悦的平衡之感。含铅玻璃和优雅的拱门是以石灰石为框架的主入口的特色，它将人引至静谧的休息厅。住宅前室以工匠建造的法式门为亮点，法式门通向用于娱乐的入口庭院。经过宴客厅和音乐厅，即可步入旋转楼梯，楼梯从地下室的影院前厅延伸至三层的游戏室。设计采用对称与平衡的经典主题，中央楼梯间的主轴向住宅后部开放。作为大厅的起居室和厨房贯穿整个住宅的后部。青铜饰品围绕的壁炉坐落在大厅的最南端。透过起居室后部宽大的窗户能够看到长满乔木的阳台，同时可以瞥见稍远处的草坪。厨房坐落在起居室对面，其中的一处厨房操作台足以容纳五人舒适地用餐。

三间卧室和一间主人休息室位于二层。客房设有独立的卫浴，但是前面的两间卧室需要共享一间浴室。筒形穹顶的枝形吊灯是共享浴室的设计亮点，该浴室以前入口上方的含铅玻璃窗为中心。主人休息室位于住宅后部，可俯视庭院。该住宅设备齐全，具有隐秘性和舒适性，既可以享受壁炉前的轻松时光，又可在私人露台欣赏落日美景。主人浴室极尽奢华，落地式浴盆和超大型蒸汽淋浴并用。

住宅的楼层平面设计简约、优雅。住宅围绕中央旋转楼梯进行设计，充分利用了建筑的面积。房间外部的尺寸和内部的尺寸都营造出了舒适的宜居空间。除了利用现代化的家用设备，该住宅也体现了法国人享受生活赐予的激情的理念。

Inspired by the timeless appeal of the traditional architecture of the French Renaissance, the Sunset house has stately brick walls with carved limestone accents to convey a sense of permanence and importance. A limestone belt course divides the first and second stories; the detail is elevated at the front entry to establish a dominant hierarchy. With steeply pitched roofs and flares at the eave, the symmetrical facade creates a pleasing balance as you approach the house. The main entrance, distinguished by elegant arched doors with leaded glass and framed in limestone, leads to an intimate foyer. The front rooms of this home boast craftsmen built French doors leading to the entry court for entertaining. Passing by the formal dining and music rooms, guests are lead to the circular stairs which stretch from the theater vestibule in the basement up to the third floor game room. Classical motifs of symmetry and balance are celebrated as the main axis of the central stair hall opens to the rear of the home. The great room and kitchen function as a grand hall stretching across the back of the home. An ornamental bronze surround finishes off the fireplace which anchors the south end of the great room. An expanse of windows along the rear of the great room offers views of the arbor terrace and a glimpse of the lawn beyond. The kitchen sits opposite the great room with an island large enough to comfortably sit five people.

Three bedrooms and a master retreat are located on the second floor. The guest bedroom has its own bathroom, while the two front bedrooms share a joined bath. A barrel vault highlights the chandelier in this joined bathroom which centers on the leaded glass window above the front entry. The master retreat sits at the rear, overlooking the yard. Providing privacy and comfort, this well-appointed room provides relaxation in front of the fireplace or sunset views from the private terrace. The master bath is complete with his-and-hers vanities, a free standing tub, and an oversized steam shower.

This home has a simple, yet elegant floor plan. Designed around the central circular stair, the house completely utilizes its square footage. The scale of the house from the exterior and the rooms of the interior create a comfortable, livable space. In addition to utilizing modern amenities, this home embodies France's national passion for savoring life's rewards in the comfort of a pleasingly appointed home.

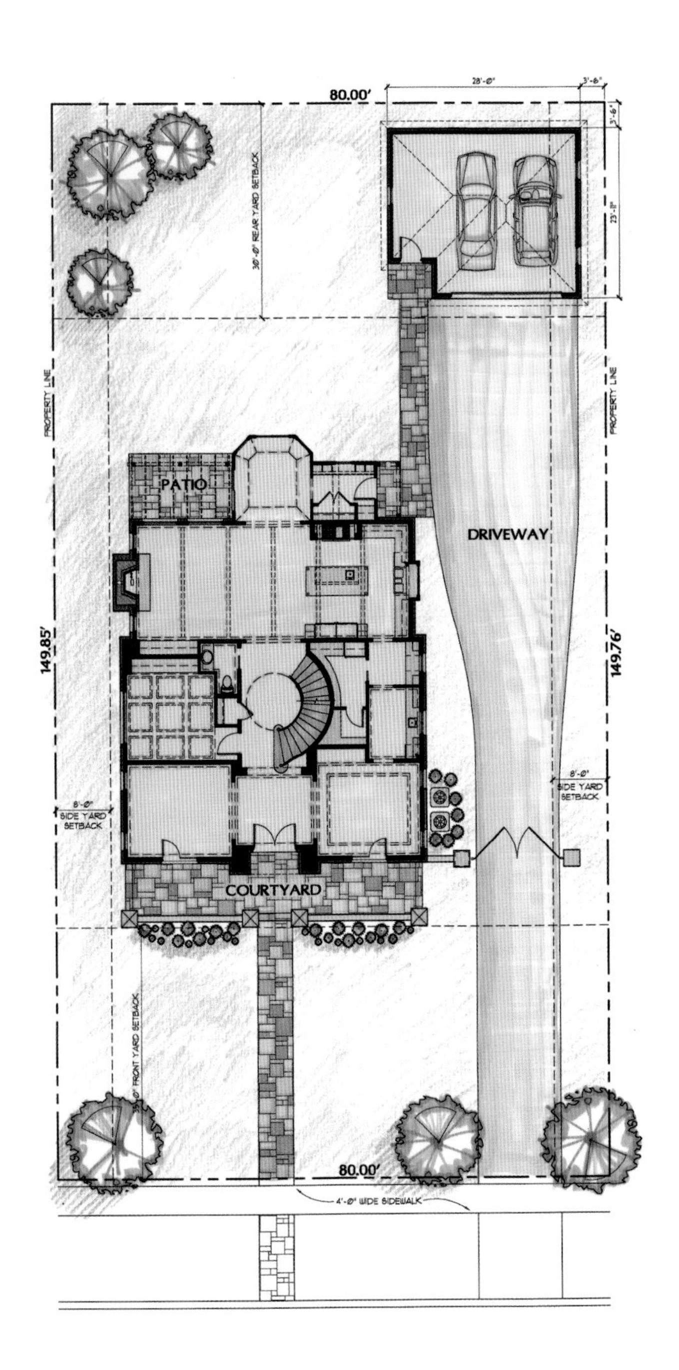

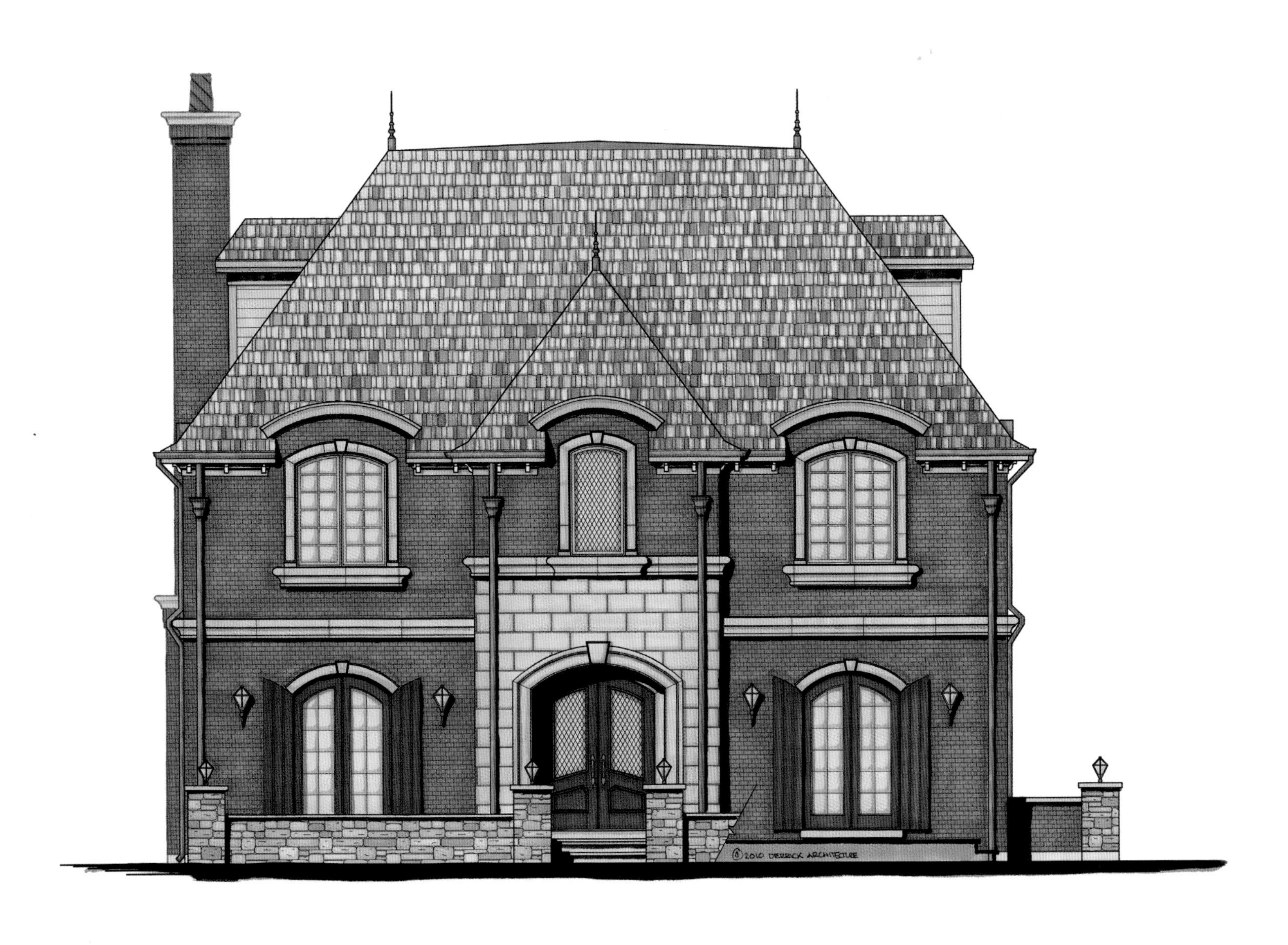
©2010 DERRICK ARCHITECTURE

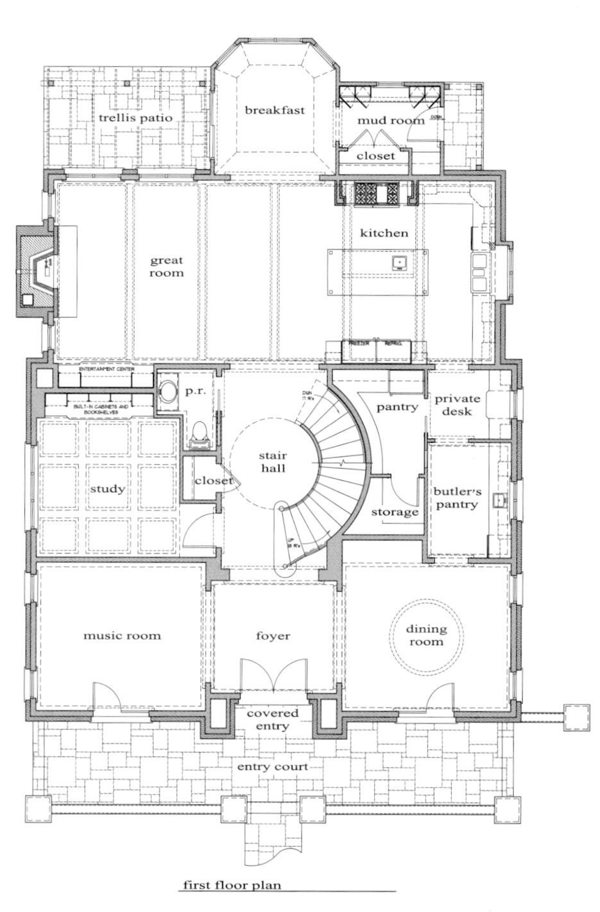

first floor plan

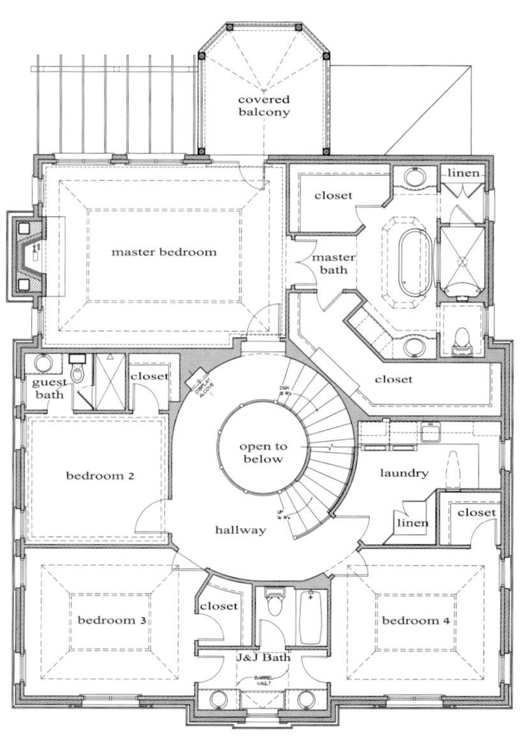

second floor plan

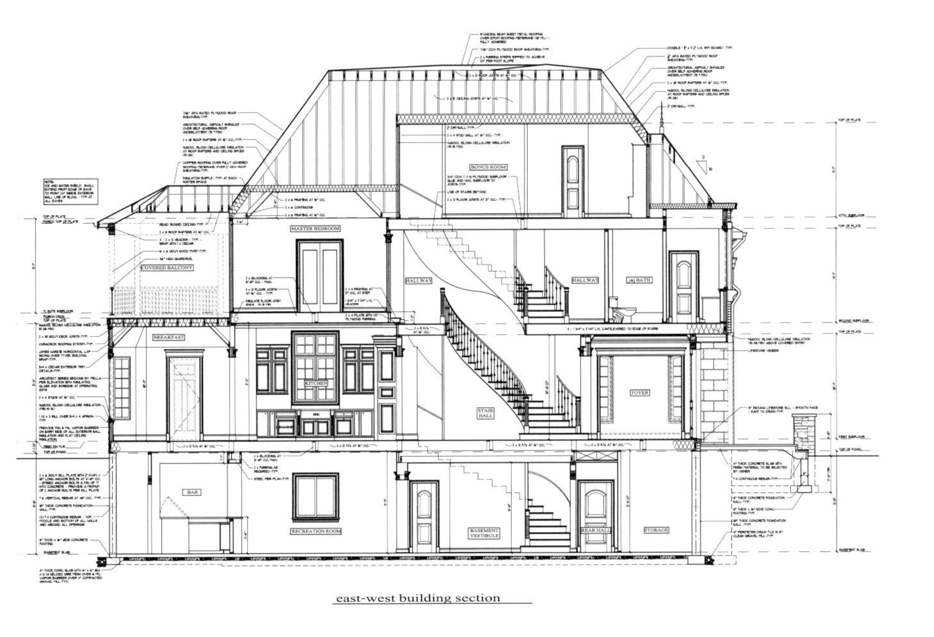

east-west building section

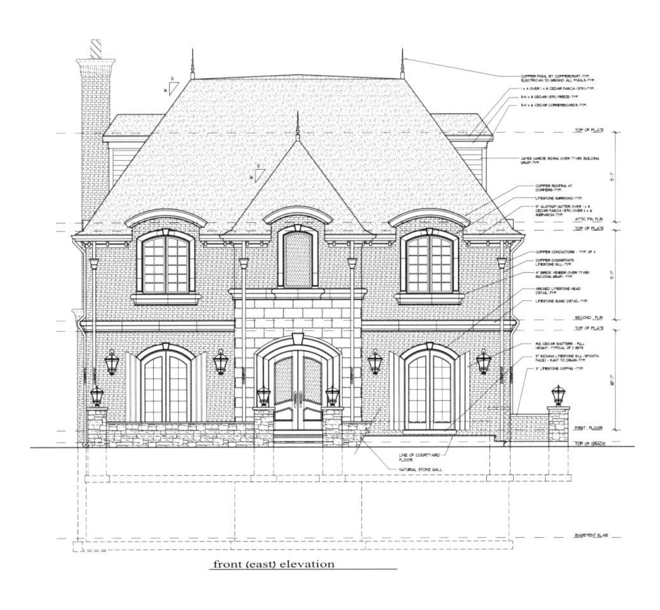

front (east) elevation

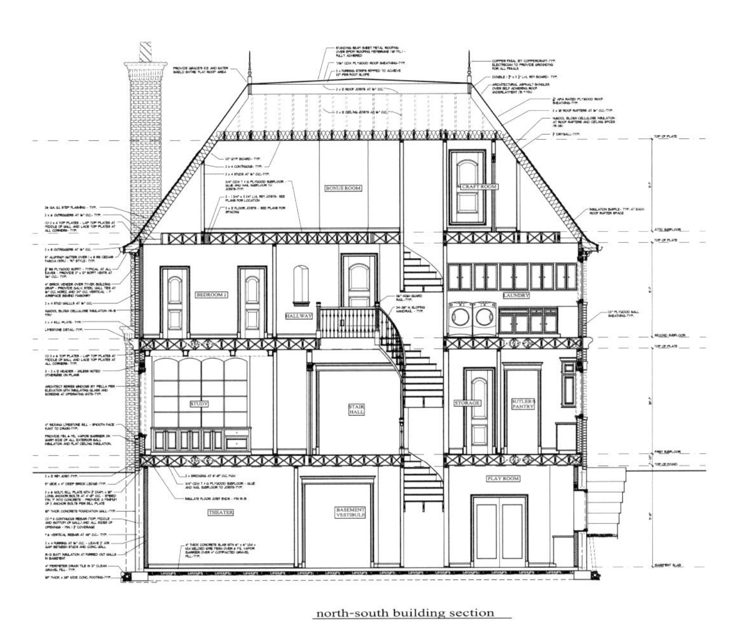

north-south building section

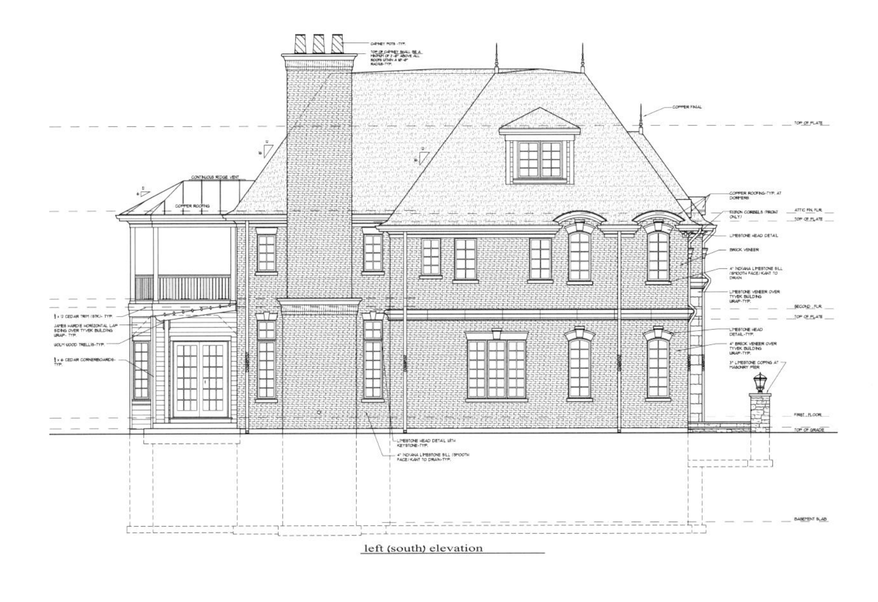

left (south) elevation

Domani 建筑与规划事务所

Domani Architecture + Planning Inc.

Fabio Rigo de Righi 是 Domani 建筑与规划事务所的创始人，意大利的成长经历和丰富的旅行经历使他受到多种建筑环境的影响。从城市环境到最私密的空间，对细节的全方位追求是他前进的动力。三十多年的设计和建筑经验使他对建筑环境拥有敏锐的感觉，这是多数人所不具备的。

年轻时期频繁的旅行以及他对文化和环境的激情的自我实现都驱使他投身建筑。成年的 Fabio 经常游走于美国和意大利之间。科摩、马焦雷湖、加尔达湖、米兰、罗马、佛罗伦萨，他与意大利家人的夏天经常在这些城市度过。他夏天在意大利工作，在美国接受教育。

该事务所提供的服务水准和诚信一定是你未曾感受过的。他的公司所承担的项目绝不仅仅是通过或"简单加工"过的，就像工厂的流水线一样。每一位甲方都是独特的，每一所住宅都由建筑师亲自设计，每一处细节都由建筑师亲自过目。

还有什么其他方式能够建成伟大的建筑吗？尽管这是一门代代相传的技术，但 Fabio Rigo de Righi 却只有一个。

Defined by his upbringing in Italy and many travels Fabio Rigo de Righi is driven by the built environment on all scales. He is driven by his passion for detail at all scales, from urban to the most intimate of spaces. Over 30 years experienced rafting and building have given Fabio a sense of the built environment that few others have.

Constant travel as a youth and self fulfillment of his passion for culture and environment drove him righ towards architecture. Growing up Fabio traveled internationally between the United States and Italy, specifically, Como, Maggiore, Garda, Milan, Rome, and Florence, Fabio spent his summers with his established Italian family. His summers were spent working in Italy while achieving education in the United States.

The level of service and integrity provided by this Architect is beyond that which you will receive with few exceptions. This is not an architecture firm where your project is passed along and"processed"much in the way manufacturing works. Every client is individual, every home will be designed by the architect, and every detail will be overseen by the architect.

Is there any other way to do great Architecture? After all this is a skill passed along through the generations and there is only one Fabio Rigo de Righi.

里维埃拉住宅
Villa Riviera

里维埃拉住宅
Villa Riviera

拉荷亚海岸
La Jolla Shores

这是一座西班牙风情的海滨寓所，地处著名的拉荷亚海岸社区。这一公寓是将灯光、体量、房间比例相结合的完美力作，打造出温馨的娱乐空间，与宜人的室外环境相联系。寓所后院更是享受南加利福尼亚气候的完美之地，比萨烤炉、泳池、温泉、篝火，加之绿草茵茵，可供儿童玩耍。

入口的表现风格以对来访客人的欢迎和对主人隐私的尊重为基础，所以它的体积和面积都不大。访客首先经过带大门的入口庭院，接着通过前门进入具有正式氛围的入口休息室。

房间的安排创造了主客间的理想分离。大人和儿童的娱乐室在空间上是分开的，但彼此之间的视线是相通的。

单一楼层的设计方便有小孩的住户的日常使用，同样也可以满足空巢老人的需求，是十分明智的选择。事实上，这一住宅是建在一片宽阔且较为水平的混凝土地面上的，单一的楼层设计使得建筑具有较大的占地面积。住宅沿外部庭院而建，从而塑造出了室外房间。

该住宅将真实的历史存在感与现代家庭的时尚潮流完美地融合。

It is a Spanish Colonial inspired beach retreat in the prestigious La Jolla Shores neighborhood. This home exemplifies a masterful use of light, volume, and room proportion resulting in a warm entertainment space well connected to outdoor amenities. Pizza oven, swimming pool, spa, fire pit, putting green and ample lawn for the children, the back yard of this home is perfect for the Southern California climate.

The entry statement, though not grand in volume or size, greets the visitor in a manner which respects the privacy of the homeowners. First one arrives at an entry courtyard with gates, and then one is led to the front door and entry foyer which has a formal ambiance. Once past the entry foyer a truly hospitable environment designed for family entertaining is revealed.

The arrangement of rooms creates an ideal separation between "served and service". There are inviting rooms for both adult and child entertaining, separate yet visually connected.

A single story plan is really a wise choice for day to day use of a residence with young children as well as one, which accommodates the aging empty nesters. In fact the home is on a large and fairly level building pad and one story makes for a larger footprint that is designed around exterior courtyards and literally carved exterior rooms.

This home combines a true historic presence with a contemporary flow for the modern family.

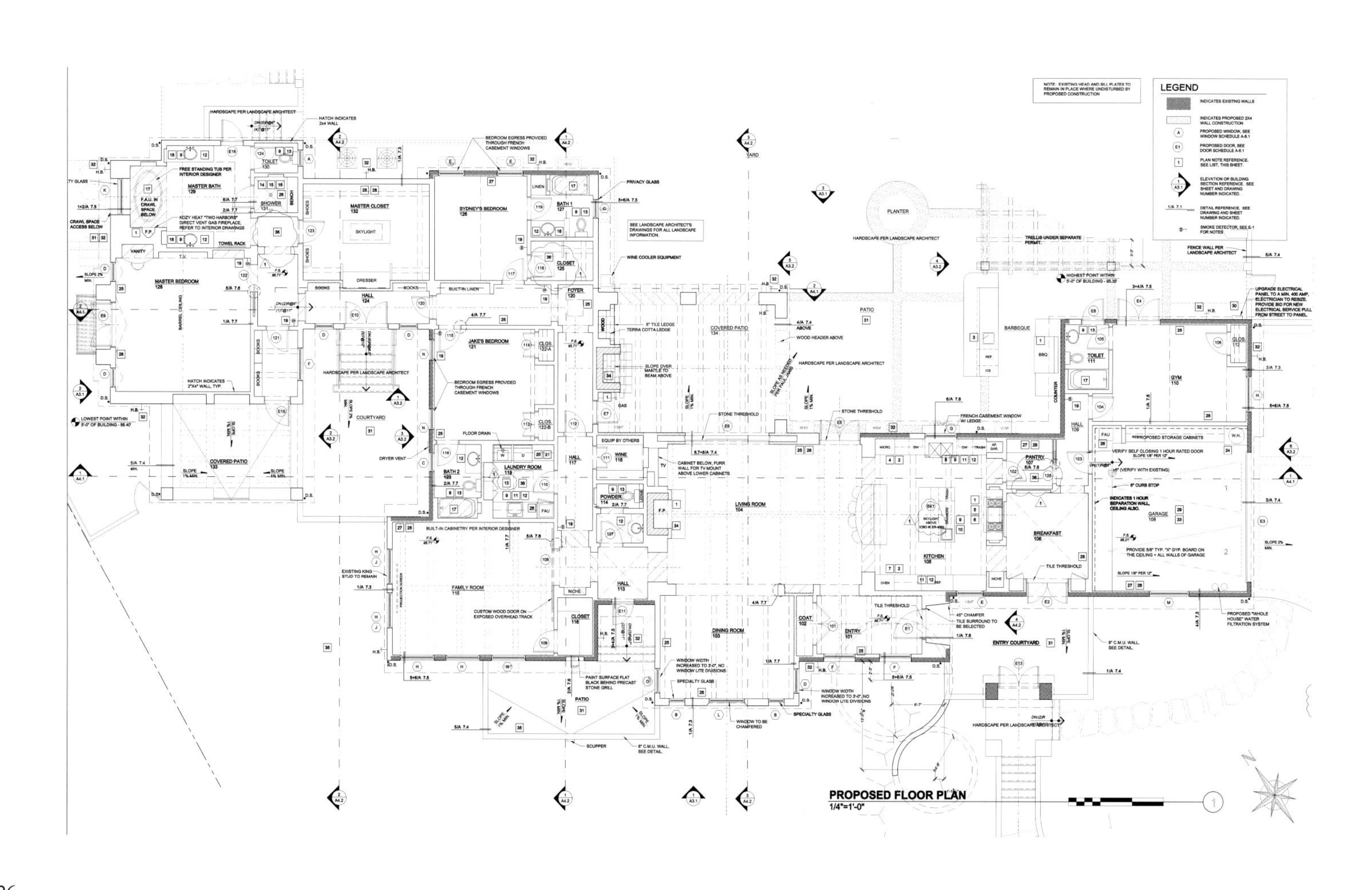
LEGEND
MASTER BEDROOM
MASTER CLOSET
MASTER BATH
SYDNEY'S BEDROOM
JAKE'S BEDROOM
COURTYARD
COVERED PATIO
COVERED PATIO
PATIO
LIVING ROOM
FAMILY ROOM
DINING ROOM
KITCHEN
BREAKFAST
GARAGE
GYM
BARBEQUE
ENTRY COURTYARD
PLANTER
PROPOSED FLOOR PLAN
1/4"=1'-0"

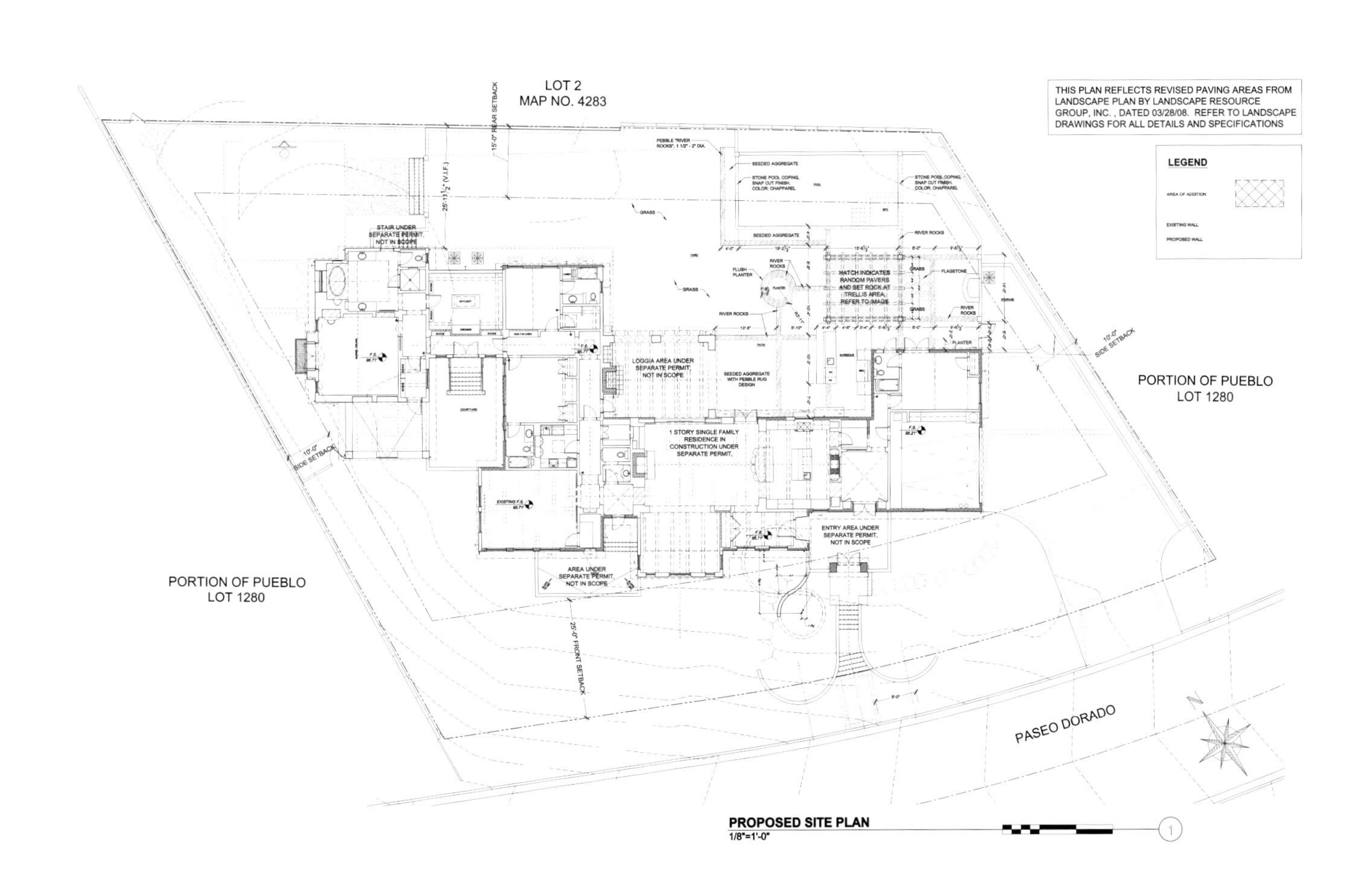
LOT 2
MAP NO. 4283
THIS PLAN REFLECTS REVISED PAVING AREAS FROM LANDSCAPE PLAN BY LANDSCAPE RESOURCE GROUP, INC. , DATED 03/28/08. REFER TO LANDSCAPE DRAWINGS FOR ALL DETAILS AND SPECIFICATIONS
LEGEND
STAIR UNDER SEPARATE PERMIT, NOT IN SCOPE
MATCH INDICATES RANDOM PAVERS AND SET ROCK AT TRELLIS AREA. REFER TO IMAGE
LOGGIA AREA UNDER SEPARATE PERMIT, NOT IN SCOPE
1 STORY SINGLE FAMILY RESIDENCE IN CONSTRUCTION UNDER SEPARATE PERMIT.
ENTRY AREA UNDER SEPARATE PERMIT, NOT IN SCOPE
AREA UNDER SEPARATE PERMIT, NOT IN SCOPE
PORTION OF PUEBLO
LOT 1280
PORTION OF PUEBLO
LOT 1280
PASEO DORADO
PROPOSED SITE PLAN
1/8"=1'-0"

Fairfax & Sammons 建筑设计公司

Fairfax & Sammons Architects P.C.

Fairfax & Sammons 建筑设计公司已成立二十余年，并荣获多项殊荣，并在纽约具有历史意义的格林威治镇和弗罗里达的棕榈滩均设有工作室。

该公司为各种建筑提供服务，包括城市建筑、城市规划、住宅设计。Fairfax & Sammons 一贯遵循传统与可持续的建筑理念，并为现代建筑的问题提供解决方案，其设计既能与自然和环境相适应，同时又根据甲方的不同需求量体裁衣。他们的建筑作品反映了均衡理论，并在几代人的学习和实践中传承下来。

公司负责人 Anne Fairfax 和 Richard Sammons 都是学术团体以及学院（例如圣母大学、耶鲁大学、格鲁吉亚理工学院和古典艺术与建筑学院）中炙手可热的评论家、教师和陪审员。

Anne Fairfax 设计过很多传统风格的建筑，并修复过多座具有历史意义的建筑，这些建筑分布在不同的城市中，如巴巴多斯、弗吉尼亚、火奴鲁鲁，她的童年时代是在这座城市中度过的，而且她在搬到纽约前一直在该市从事建筑事业。Fairfax 女士是古典建筑与经典美国学院以及皇家橡树基金会董事会的一员，并任教于查尔斯王子基金会的美国夏日学校，同时是佐治亚理工学院的哈里森访问学者。如今她还是该公司的董事长，并担任执行合伙人一职。

Richard Sammons 在传统设计方面有着十分丰富的背景，是一位在建筑领域受到国际认可的专家，任教于伦敦的威尔士王子建筑学院、纽约的普瑞特艺术学院和罗马的圣母大学。他曾为 C Howard Walker 所著的《装饰线条的理论》一书的再版编写前序，另外他最近出版的《让你的房子设计得恰到好处》一书则由查尔斯王子亲自作序。Sammons 先生在纽约的 David Anthony Easton 工作室开始了他的职业生涯，并在那里工作了很多年。他是古典建筑和经典美国学院的创始人。Richard 现任 Fairfax & Sammons 建筑设计公司的副董事长，并担任首席设计师一职。

Anne Fairfax 和 Richard Sammons 既是夫妻又是事业上的伙伴，他们携手创建了 Fairfax & Sammons 建筑设计公司，现有员工 36 人。他们也是 ICTP 的成员，ICTP 是一个由威尔士王子资助从事传统建筑工作者的国际性专业团体。他们获奖无数，其作品也被刊登在诸多杂志和文摘上，如《纽约时报》《乡村生活》《建筑文摘》《南方风格》《老房子期刊》《时期住宅》《传统建筑》以及《新旧式住宅》。他们的专题著作《美式住宅：Fairfax & Sammons 作品集》由 Rizzoli 国际出版社发行。

Fairfax & Sammons Architects P.C. established over twenty years ago, is an award-winning architectural design firm with offices in both New York' s historic Greenwich Village and Palm Beach, Florida.

Fairfax & Sammons offers the full scope of architectural services including urban, civic design, and residential design. The firm is committed to an architecture of tradition and sustainability that offers solutions to contemporary issues, incorporating the belief that buildings be designed appropriately for climate and setting and be tailored to the specific needs of the client. Their body of work reflects theories of proportion and order that have been passed down through scholarship and practice for generations.

The principals, Anne Fairfax and Richard Sammons are both sought-after critics, teachers and jurors in the academic community, at such institutions as the University of Notre Dame, Yale, Georgia Institute of Technology and the Institute of Classical Art and Architecture.

Anne Fairfax has deigned many new buildings in traditional styles, restored numerous historic houses in diverse locations such as Barbados, Virginia and Honolulu, the city where she spent her childhood and maintained an architectural practice before moving to New York. Ms. Fairfax serves on the board of the Institute of Classical Architecture & Classical America, and the Royal Oak Foundation. She taught at the Prince' s Foundation American Summer School and was the Harrison Visiting Scholar at Georgia Institute of Technology. Today she is the President of the firm, serving as managing partner.

Richard Sammons has a rich background in traditional period design and is an international recognized expert in the field of architectural proportion, having taught at The Prince of Wales' Institute of Architecture in London, Pratt Institute in New York and The University of Notre Dame in Rome. His contributions include the foreword for the re-publication of the book *The Theory of Mouldings* by C. Howard Walker. He contributed to the recently published, *Get Your House Right* with a foreword written by Prince Charles. Mr. Sammons began his career in the office of David Anthony Easton in New York, where he worked for several years, and is founding director of The Institute of Classical Architecture & Classical America. Richard is the Vice President of the firm, serving as chief designer.

Partners in business as well as in marriage, Anne Fairfax and Richard Sammons are founding partners of the firm, which today comprises of thirty-six members. They are both members of the INTBAU College of Traditional Practitioners (ICTP), an international professional body for practitioners in traditional architecture, under the auspices of His Royal Highness, the Prince of Wales. They are the recipients of numerous awards and their works has been featured in *The New York Times, Country Life, Architectural Digest, Southern Accents, Old House Journal, Period Homes, Traditional building* and *New Old House*. Their monograph, *American Houses: The Architecture of Fairfax & Sammons* is published by Rizzoli International.

作　品

Selected Work

20 世纪 20 年代乔治王时代的砖质复兴住宅
1920's Brick Georgian Revival House

新乔治王时代住宅
New Georgian Residence

文艺复兴住宅
Renaissance Revival Estate

木板式复兴住宅
Shingle Style Restoration

Breakers 联排公寓
Breakers Row Apartment

哥特式住宅
Gothic Style House

新联邦式住宅
New Federal House

摄政风格建筑
Regency Style Estate

20 世纪 20 年代乔治王时代的砖质复兴住宅

1920's Brick Georgian Revival House

肯塔基州列克星敦市

Lexington, Kentucky

20世纪20年代乔治王时代的砖质复兴住宅位于肯塔基州的列克星敦市。业主要求新建主人套房、书房、后院门廊、池塘，同时美化景观，并重新设计前门入口和楼梯。他们也一直在寻求一座理想的住宅来放置那套 19 世纪的美式家具以及装饰性艺术品。

通过将旧建筑与新建筑融合在一起，以打造一处天衣无缝的扩建项目。同时，还将住宅的所有功能进行更新，并增建了一些内部建筑

该建筑设计有一种独特的砖石饰面——该饰面混合使用了三种颜色的砖石，它们都经过了酸洗和喷砂处理。若将增建部分的砖块与原有砖块相匹配是极度困难的。由于后立面的式样，建筑师可在新的早餐室上方加盖一种内嵌式圆屋顶的弓形屋顶，使这间没有任何窗户的早餐室充满阳光。后立面增建了一个新的门廊和凉廊，从而使厨房和图书室更加和谐统一。

门廊、新酒窖、嵌板式图书室采用新的预制木质建筑材料和抹灰工艺，“长臂猿”房间采用黑橡木色的抹灰天花板，起居室、餐厅和楼梯间天花板则采用装饰石膏线条，并全部采用新模制成型工艺。

The owners of the 1920's brick Georgian revival house, in Lexington, Kentucky required a new master suite, study, back porches, pool and landscaping, as well as a redesign of the front entrance and the stair. They were also looking for a proper setting for their eclectic collection of 19th century American furniture and decorative arts.

To blend the old with the new to create a seamless addition, while updating all features of the house and adding interior architecture.

There is a unique brick finish – a combination of three colors of brick, which had been acid-washed and grit blasted. Matching the existing brick for the addition was extremely difficult. A cut-out in the rear facade allowed architects to install a saucer-dome over the new breakfast room with a built-in cupola. This allowed light into the breakfast room where no other window existed before. A new rear porch and loggia was added to the rear facade to unify the kitchen wing with the library wing.

New millwork and plaster work included the front portico, a new wine cellar, a paneled library, addition of a Jacobean plaster ceiling to the "Gibbons" room and decorative plaster mouldings to the ceilings of the living room, dining room and stair hall and complete new moulding program throughout.

MUDROOM
PANTRY
VESTIBULE
REAR PATIO
LOGGIA
KITCHEN
BREAKFAST RM
LIBRARY
BUTLER'S PANTRY
STUDY
DINING ROOM
ENTRY HALL
LIVING ROOM
SUN ROOM
BEDROOM #3
MASTER BEDROOM
LANDING
VESTIBULE
MASTER BATH
BATH
HER DRESSING CLOSET
STAIR HALL
VESTIBULE
HER NEW CLOSET
HER NEW DRESSING ROOM
BEDROOM #2
LIBRARY
HIS REVISED DRESSING ROOM
BALCONY

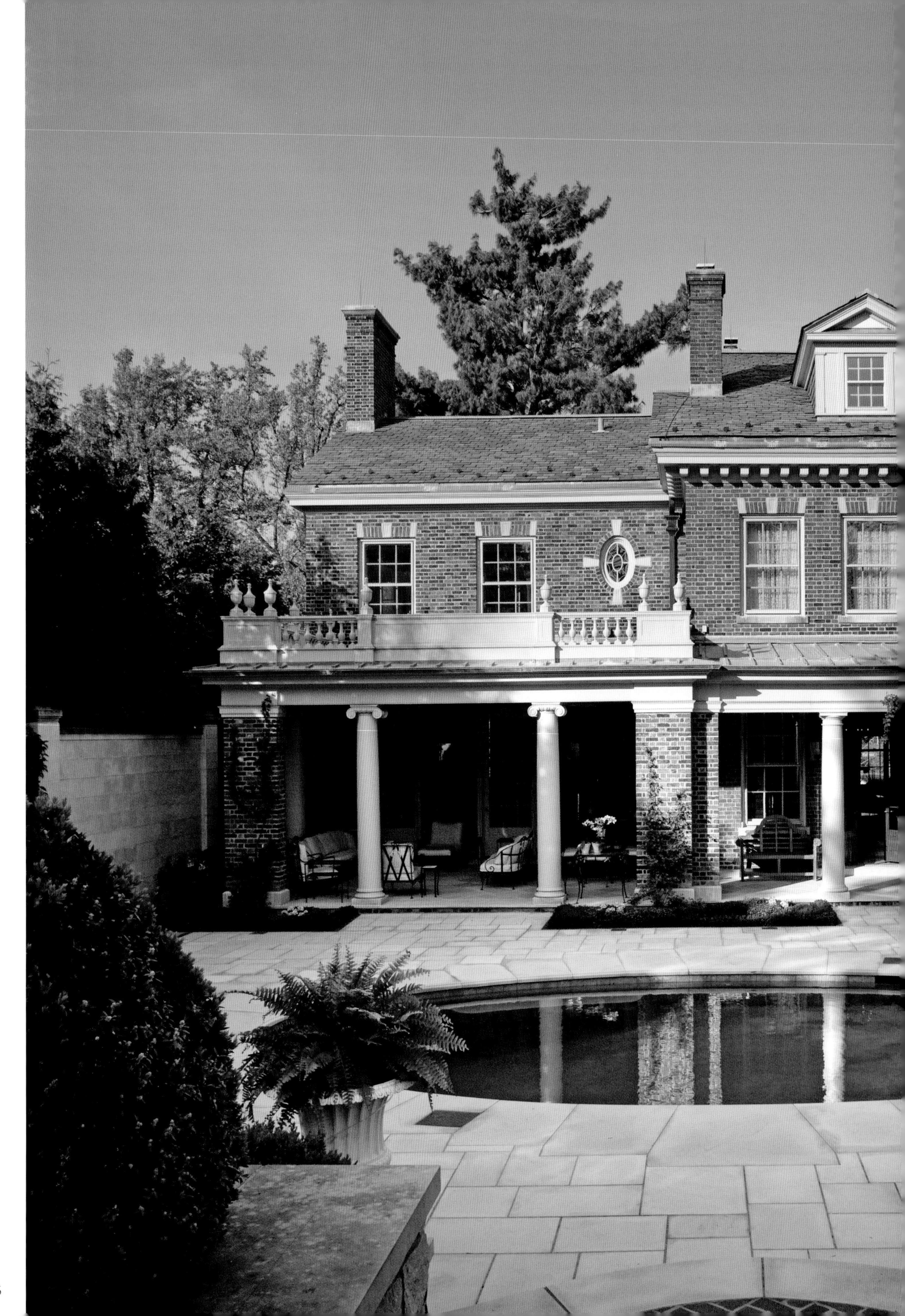

新乔治王时代住宅
New Georgian Residence

康涅狄格州华盛顿市
Washington, Connecticut

这座小巧却引人注目的住宅坐落于美国康乃狄克州西北部的一个孤立的山顶上。由于该建筑地处崎岖险峻之地，因此在设计建筑特点时，建筑师会参考一些苏格兰先例，项目的细节处理和比例方面的灵感均来自于苏格兰具有帕拉迪奥建筑风格的威廉·亚当，他是《Vitruvius Scotticus》的作者，也是更著名的建筑师罗伯特和詹姆斯的父亲。亚当的作品中充满了粗野的、后巴洛克式的繁荣，从而使建筑可以使住宅被建造在景观之中，同时还彰显了业主对古典建筑语言的兴趣和渊博学识。建筑材料也根据其纹理而精挑细选：手工制作的露头砖与精细的石质檐口、烟囱帽和尖顶形成鲜明对比。建筑室内，带有嵌板的房间可远观康乃狄克州的山峦。该项目还将花园一直延伸至远处的景观中。尽管住宅相对较小（不足 325m^2），但是经典的比例与坚固的细节处理创建了一个整体效果远大于其各个部分效果累加的建筑。

This small, dramatically architectural house stands on an isolated hilltop in northwestern Connecticut. The ruggedness of the site led architects to look at Scottish precedent in establishing the character of the design, and much of the detailing and proportioning are inspired by the Scottish Palladian architect William Adam, author of *Vitruvius Scotticus,* and father of the more famous Robert and James. The robust, late baroque flourishes found in Adam's work allows the house to establish its presence in the landscape while providing a sophisticated expression of the owner's own knowledgeable interest in the classical language of architecture. The materials were deliberately chosen for their texture: handmade brick with burned headers contrasting with substantially detailed stone cornices, chimney caps and finials. Inside, paneled rooms have far reaching views across the Connecticut hills. Work is currently proceeding on extending the immediate garden into the landscape beyond. Although relatively small (less than 325m^2), classical proportion and robust detailing create a composition whose the whole feels far greater than the sum of its parts.

LITCHFIELD

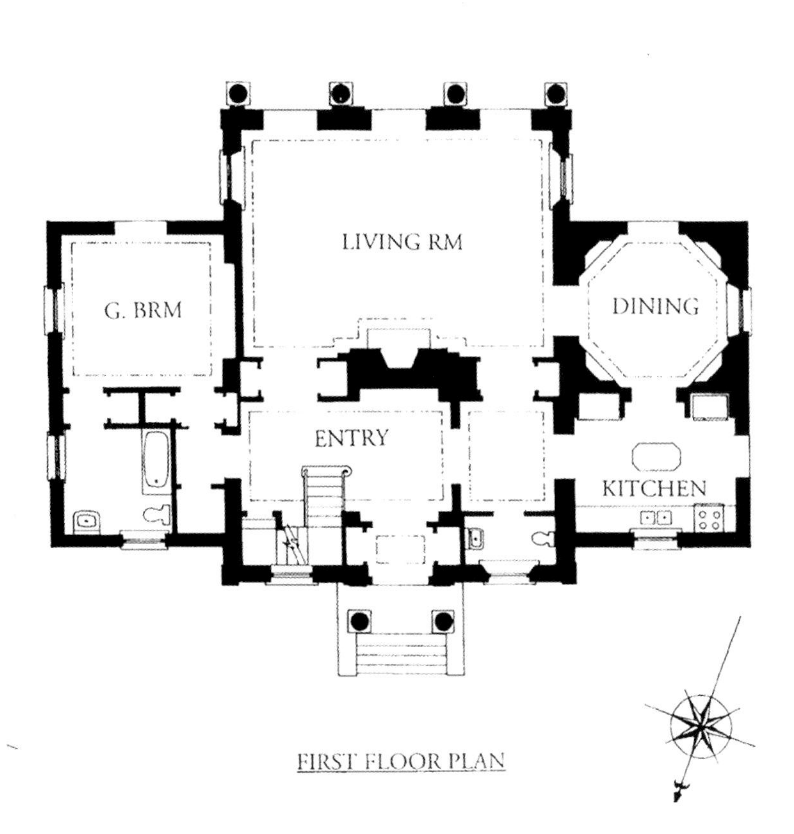

LITCHFIELD

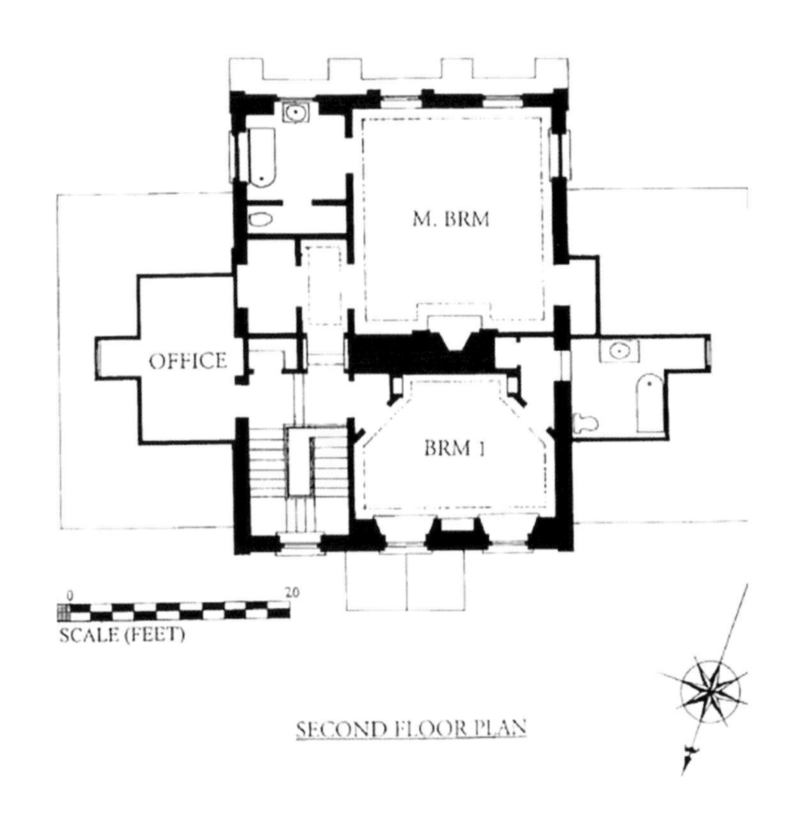

文艺复兴住宅
Renaissance Revival Estate

佛罗里达州棕榈海滩
Palm Beach, Florida

在建筑师们根据业主（一位互联网先锋）的要求，对这座位于棕榈滩、由Maurice Fatio设计的历史性住宅进行增建和整修的过程中，他们发现他们的工作既在精神上与原建筑师的意图相契合，同时又在形式上为现在的业主扩大了该住宅的功能。来自旧金山的室内设计师Rebecca Bradley和来自纽约的Brian McCarthy在工作中紧密配合，通过采用高档的装饰性壁炉架、石质浮雕、刷漆的木质天花板以及大量的装饰性饰面，室内的饰面质量得以提高和修整。

27间服务性卧室由一个新增建的侧翼所代替，它可以同时满足非正式的居住和用餐需求。另外，还新增了一座观光塔、一个186m^{2}的地下石灰岩圆顶酒窖，同时还在主卧中增设了一个凉廊以及一个摩洛哥式花房。此外，沃斯湖畔还建有一个全新的艇库和一座海滨别墅，该别墅由一条地下隧道与主建筑住宅相连。住宅所有的石头均用石灰岩和多米尼加共和国的贝壳灰岩与珊瑚石所代替，对这些石材的雕刻和打磨均在英国和当地完成。住宅的总面积超过5574m^2。该项目的建筑设计师为Fairfax & Sammons建筑设计公司，景观建筑师为Robert Truskowsky。

Adding to and renovating an existing historic residence by Maurice Fatio in Palm Beach, for the owner, an internet pioneer, architects found themselves working both within the context of the original architect's intention as well as expanding the use of the house for the present owner. Working in close conjunction with the interior decorators, Rebecca Bradley of San Francisco and Brian McCarthy of New York, the quality of the interior finishes were improved and restored and included the addition of museum-quality decorative mantels, stone reliefs painted wood ceilings, and a myriad of decorative finishes.

Twenty-seven service bedrooms were replaced by a complete new wing addition that incorporated the need for informal living and dining. A viewing tower was added as well as 186m^2 underground limestone vaulted wine cellar, a new loggia to the master bedroom, and a Moroccan garden room. In addition, there is a complete new boathouse on the Lake Worth side and a new beach house on the ocean side, which is connected to the main property by a tunnel under the road. All stone on the house was replaced with limestone and Dominican Republic coquina, or coral stone, carved and milled both locally and in England. The sum total of the house is over 5,574m^2. Fairfax & Sammons Architects P.C. was the design architect for the project, and the landscape architect was Robert Truskowsky.

木板式复兴住宅
Shingle Style Restoration

康涅狄格州格林威治
Greenwich, Connecticut

现任业主发现了这座被遗弃的住宅，并完全剥去了它不合时宜的现代田园式“外衣”，将其重建为一座木板式乡村住宅。该项目坐落于地广人稀的格林威治的内陆地区，业主以低于预算的价格购得了这座2044m^{2}的住宅，并迅速对其进行设计和重建。板式的外表将会使主要房间内充满阳光、比例匀称、细节精巧。主入口通往两层高的橡木大厅和塔楼，日光从塔楼射入室内，照亮了石灰岩地面。尽管房屋较大，但拱形和方格天花板、炉边与窗边的座位等细节方面都为房间赋予了一种亲密感。超大的公共娱乐室位于该住宅的一层，向景观庭园和门廊开放，这是对该项目的完美补充，同时也增加了私人乡村寓所的私密性。新建的游泳室和网球馆使住宅更加完善。

Found abandoned and derelict by its present owners, this residence was completely stripped of its awkward, modern, prairie-style garments and remade as a Shingle Style country house. Located in substantial acreage in upcountry Greenwich, this 2,044m^2 house was brought in under budget and was designed and constructed on a fast-track schedule. The shingled exterior gives way to a light-filled series of principal rooms, classically proportioned and detailed. The main entrance opens into a two-story oak hall and stair tower from which light streams, illuminating the limestone floor. Despite its grand scale, details such as vaulted and coffered ceilings, inglenooks and window seats impart an intimacy to the rooms. A house for entertaining on a large scale, the public areas on the ground floor open up to landscaped grounds and terraces, which complement the residence and contribute to the feeling of a private country retreat. A new pool house and tennis pavilion complete the estate.

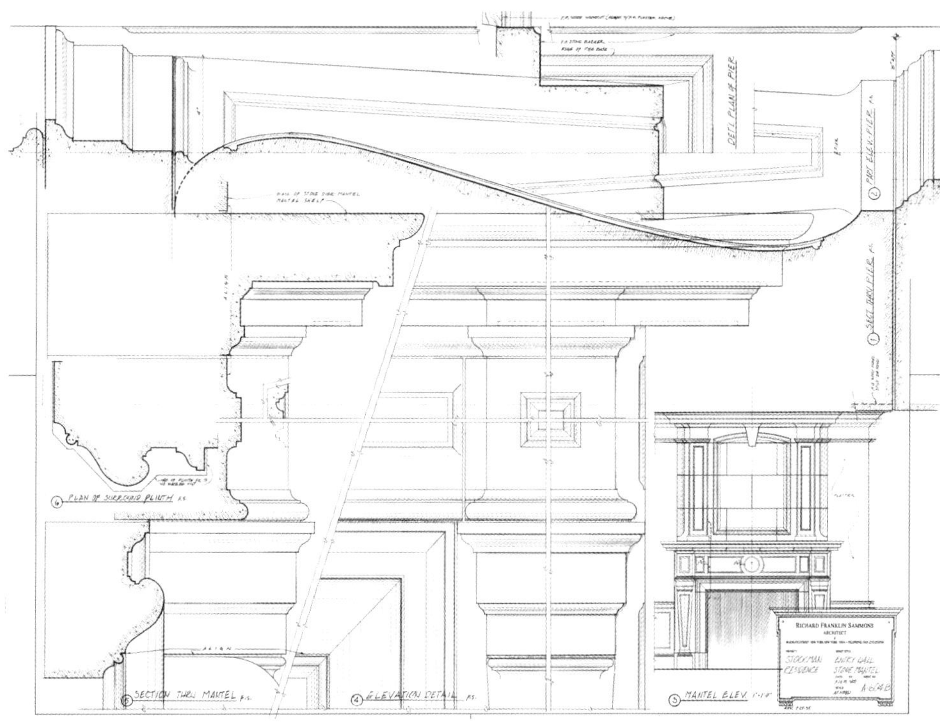

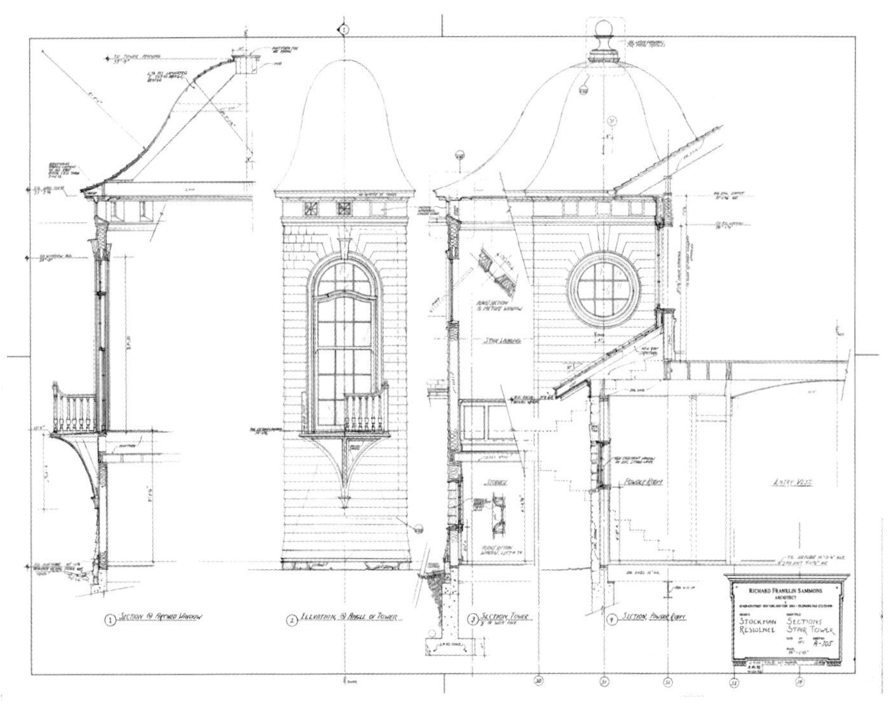
RICHARD FRANKLIN SAMMONS
ARCHITECT
STOCKMAN RESIDENCE
SECTIONS STAIR TOWER
POWDER ROOM
ENTRY HALL

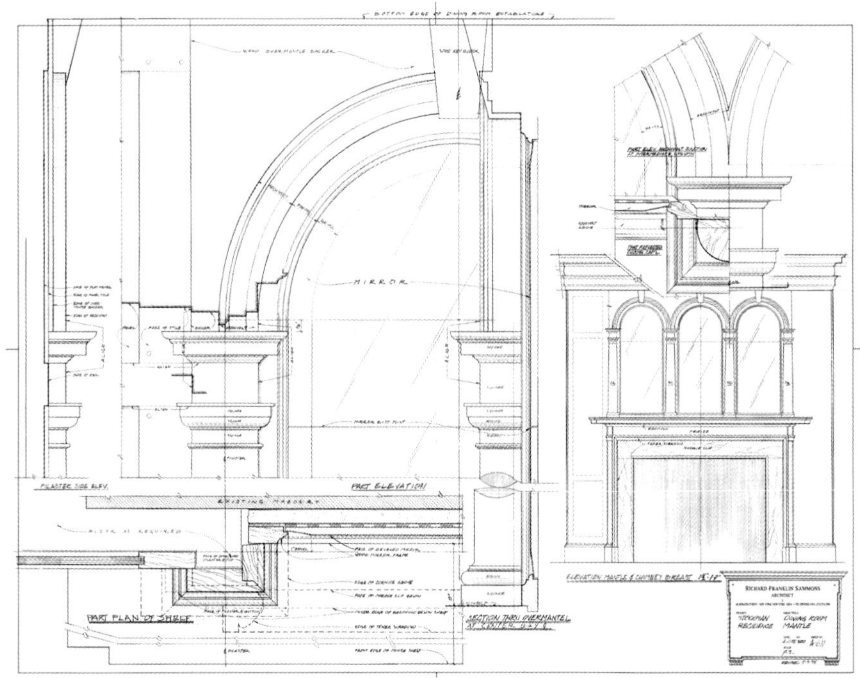
MIRROR
EXISTING MASONRY
PILASTER SIDE ELEV.
PART ELEVATION
PART PLAN OF SHEET
SECTION THRU OVERMANTEL AT CENTER LINE
RICHARD FRANKLIN SAMMONS
ARCHITECT
STOCKMAN RESIDENCE
DINING ROOM MANTLE

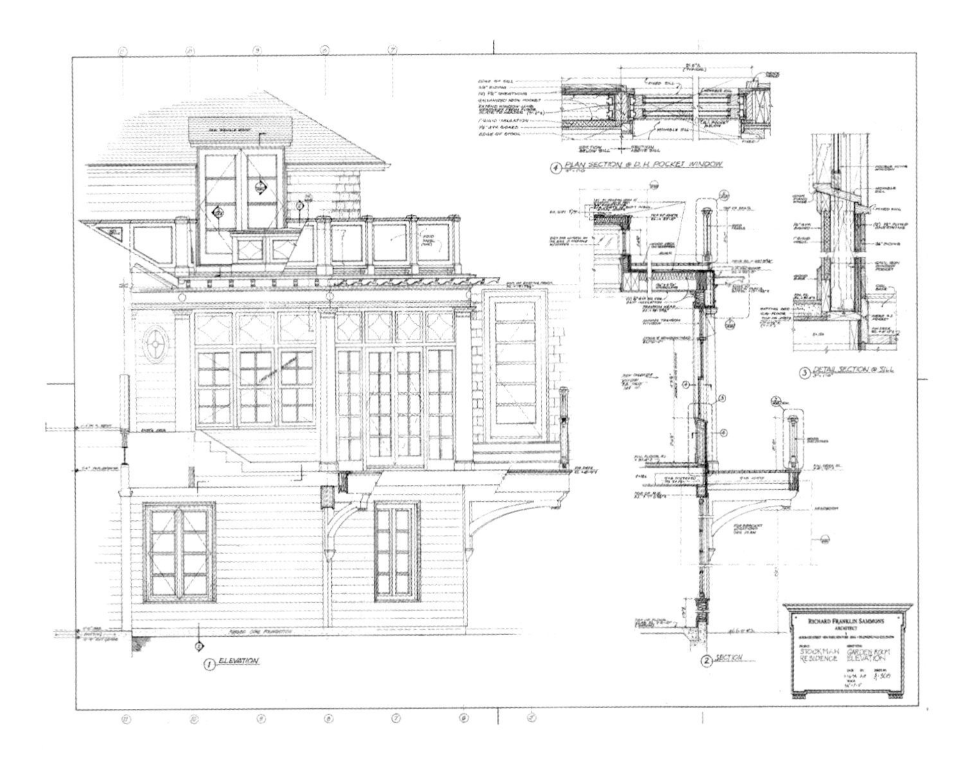
PLAN SECTION @ D.H. POCKET WINDOW
DETAIL SECTION @ SILL
ELEVATION
SECTION
RICHARD FRANKLIN SAMMONS
ARCHITECT
STOCKMAN RESIDENCE
GARDEN ROOM ELEVATION

GERHARD RICHTER

Breakers 联排公寓
Breakers Row Apartment

佛罗里达州棕榈滩
Palm Beach, Florida

该项目是为一对九十多岁的新婚夫妇设计的，这对夫妇将他们的公寓连在一起从而开始了他们的新生活。该项目位于棕榈滩著名的 Breakers 酒店的一层，目前面积已经超过 460m²，使他们的生活能够完美地融合在一起。

业主收藏了世界一流的艺术藏品以及具有博物馆收藏价值的古玩，他们希望在这栋新公寓中展出这些藏品。来自马萨诸塞州波士顿的著名室内设计师 William Hodgens 是该项目设计团队的一名主要成员。在 40 年的从业时间里，他与这两位业主都合作过，所设计的项目也都非常受欢迎，因而这些成功的经验使其能够在规定的时间内完成该项目。

This project was designed for a nonagenarian newlywed couple who were starting their new lives by combining their apartments. Located on the ground of the celebrated Breakers Hotel in Palm Beach, this condominium, now over 460m², accommodated their combined lives seamlessly.

The owners had a world class art collection as well as museum-quality antiques, which they desired to showcase in the new apartment. The well-known interior designer William Hodgens, of Boston, Massachusetts, was an integral member of the design team. Having worked with both clients before over a span of 40 years was a very welcome level of experience which allowed the project to be completed in record time.

哥特式住宅
Gothic Style House

纽约贝德福德
Bedford, New York

Fairfax & Sammons 建筑设计公司受人之托对 20 世纪 20 年代美国建筑师 Kerr Rainsford 之前的私人住宅进行翻新，这栋住宅隐藏在纽约贝德福德占地 16ha. 的公园式区域中。这栋哥特式（有时被称为“斯托克布鲁克哥特式”，是早期经济大亨们喜爱的建筑风格）住宅的大部分一直保持原貌，不过前不久，该住宅的现任业主与建筑师联系，希望能够设计一栋附加建筑，将该地产进行翻新并新建附加建筑。建筑师考虑到天马行空而又简朴的别墅建筑风格，将这栋大面积的附加建筑建造在住宅的后部，从而构成了“L”形。光线充足的全新厨房位于新老建筑的交叉点，这使得住宅能够向外开放，同时也将现有住宅逐渐过渡到全新的引人注目的家庭娱乐室。附加建筑外部所使用的材料以及在细节处理方面与现有建筑保持一致，这样便与原有建筑完美地融合在一起。家庭娱乐室是手工凿成的，用重型木结构组成，里面还设有大型的炉边石灰岩壁炉，它将外部的别墅区域补充完整。位于该基地的另外一处别墅也经过了翻新，建筑师将其改造成了一处游泳房和客房，停车场和客房位于车道入口。

摄影：Durston Saylor

Fairfax & Sammons Architects P.C. were commissioned to renovate the former private residence of the 1920's American architect Kerr Rainsford which was hidden in a 16ha. park-like setting in Bedford, New York. Until very recently the gothic style house, sometimes called "Stockbroker Gothic" for the penchant that early broker barons had for the style, remained largely unaltered until the present owner contacted the architects to design an addition, renovations and outbuildings to the estate. In deference to the whimsical, rustic cottage architecture, the large addition was placed at the rear of the residence forming an "L" shape. The light-filled new conservatory kitchen was placed at the intersection of the old and new, opening up the residence and providing a transition from the existing house into the new dramatic family room. The exterior of the addition uses the same materials and details as the existing house, allowing it to blend seamlessly with the old. The cottage scale of the exterior is complemented by the family room with its hand-hewn, heavy-timber structure and large inglenook limestone fireplace. Another cottage on the site was renovated for use as a pool house and guest quarters and a garage and guest quarters were placed at the driveway entrance.

新联邦式住宅
New Federal House

组约古柏镇
Cooperstown, New York

Fairfax & Sammons 建筑设计公司受托为 Anhauser Busch 公司创始人的外孙在纽约州北部的家族啤酒花农场设计一栋新房。这座石质新房建在祖父母的房址之上，旨在成为后代子孙的新家园。

房子坐落于著名的 Glimmerglass 湖畔，俯瞰湖光水色，美不胜收。新房由当地承重石建成，采用新联邦设计风格，并与周围的邻里住宅融合在一起 。

原来建造的木屋无法抵挡古柏镇的严冬，需要重新盖一座房子。经房主同意，一座更坚固的新房建在了原来的房址之上。房子的主体采用当地的纽约青石构筑，侧翼建筑为木质框架镶嵌以层列木瓦而成，屋顶则由重型石板覆盖。这些材料是联邦时期本土建筑的一大特色。本土建筑材料与传统建筑手法相结合，使得房子看起来坚固耐用，又与当地崎岖多岩的景致融为一体。

一踏进这座 930m^2 的房子，通向二楼的双层高走廊便会映入眼帘。弯曲的楼梯风格优美雅致，把各个楼层连接起来。走廊的另一头是客厅，客厅的左面是一间四周镶嵌松木板的图书室，右面则是一处布置井然的餐厅。

这三间主房成行排列，有很多大窗户和法式房门，可以看到房子后面的全部风景，远处则是湖上风光。

进入房子里面，嵌板房间各部分都饰有精心定制的木制品。中心门廊中设有一个悬挑式阳台，为主卧室提供了俯瞰湖景的观景廊。巴拉迪欧风格的窗户为稍显强势的建筑立面添加了一丝柔和的曲线。

侧翼通向朝南的凉廊，东面是一间客房，厨房和早餐室在西面，卧室则在楼上。早餐室镶着嵌板，厨房与之相通，其风格都不像主房那样正式，笨重的房梁支撑着天花板，其中还设有一个石刻大壁炉和一些粗木家具。

Fairfax & Sammons Architects P.C. were commissioned to design a new house for the grandson of the founder of Anhauser Busch, on the family hops farm in upstate New York. Built on the site of his grandparents' house, this new stone house was intended to become the new family home for generations to come.

Sited on the famous Glimmerglass Lake, it commands a breathtaking view across the water. The house is constructed of load bearing local stone and is designed in a Neo-Federal style, allowing the house to blend in with its neighbors.

The original wood house had not been built to withstand the harsh Cooperstown winters, and needed to be replaced. The owner agreed, and a more substantial house was built on the site of the previous one. The main core of the house is constructed with the local New York bluestone, the flanking bays are wood framed with coursed shingles, and the roof is heavy slate. These materials are a feature of local buildings of the Federal Period. The use of local materials worked in a traditional manner gives the house its sense of durability and its feeling of belonging to the rugged local landscape.

Upon entering the 930m^2 house, a two-storey entrance gallery opens up to the second floor. The curved stair provides a graceful connection from each floor. Beyond the gallery lies a living room that is flanked by a pine paneled library on the left and a formal dining room on the right.

These three principal rooms, connected by an enfilade, have numerous large windows and French doors allowing for a panoramic view from the rear of the house, with the lake in the distance.

Inside, paneled rooms were carefully detailed with fine custom millwork throughout. The central portico contains a suspended balcony providing a porch for the master bedroom overlooking the lake. Palladian windows add a soft curve to a somewhat strong and dominant facade.

In the wings, which open onto south facing loggias, there is a guest bedroom suite on the east side and the kitchen and breakfast room on the west, with bedrooms on the floor above. The kitchen and interconnecting paneled breakfast room introduce a less formal note, with a heavy beamed ceiling, a large, stone checked fireplace, and rustic furniture.

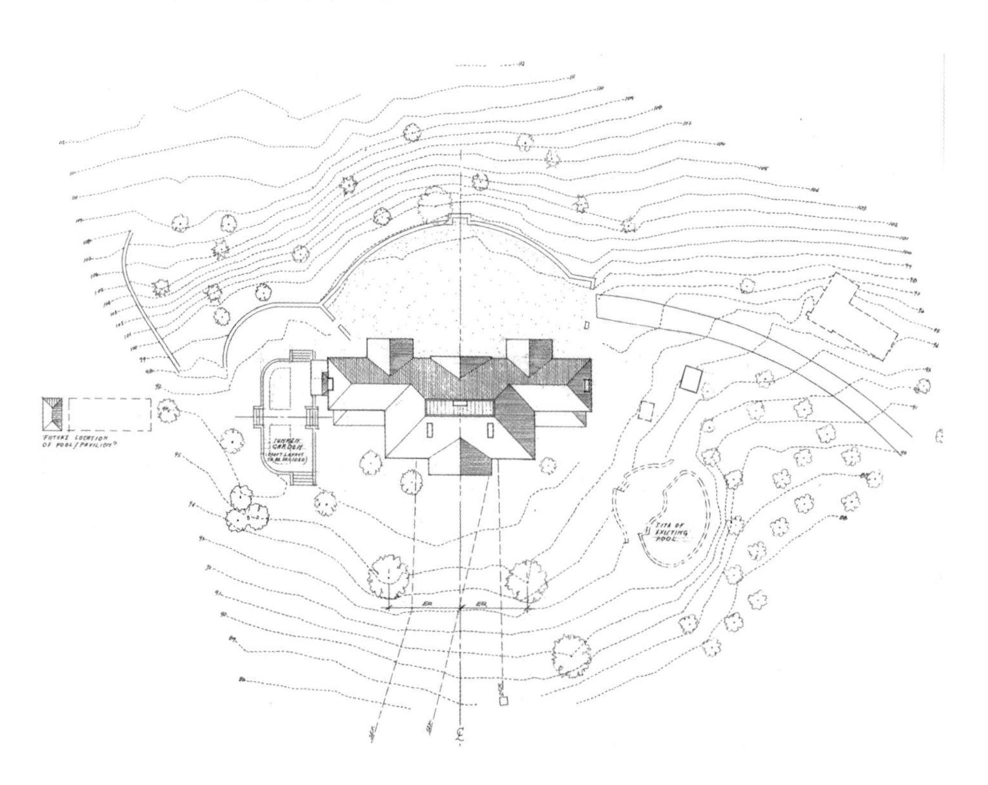

FARMLANDS

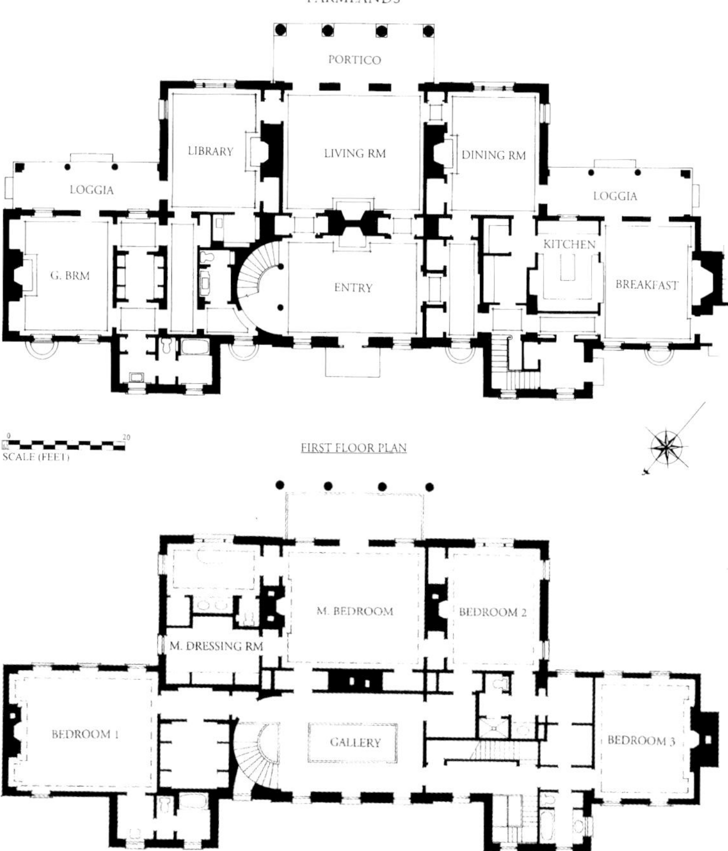
PORTICO
LIBRARY
LIVING RM
DINING RM
LOGGIA
LOGGIA
KITCHEN
G. BRM
ENTRY
BREAKFAST
SCALE (FEET)
FIRST FLOOR PLAN
M. BEDROOM
BEDROOM 2
M. DRESSING RM
BEDROOM 1
GALLERY
BEDROOM 3
SECOND FLOOR PLAN

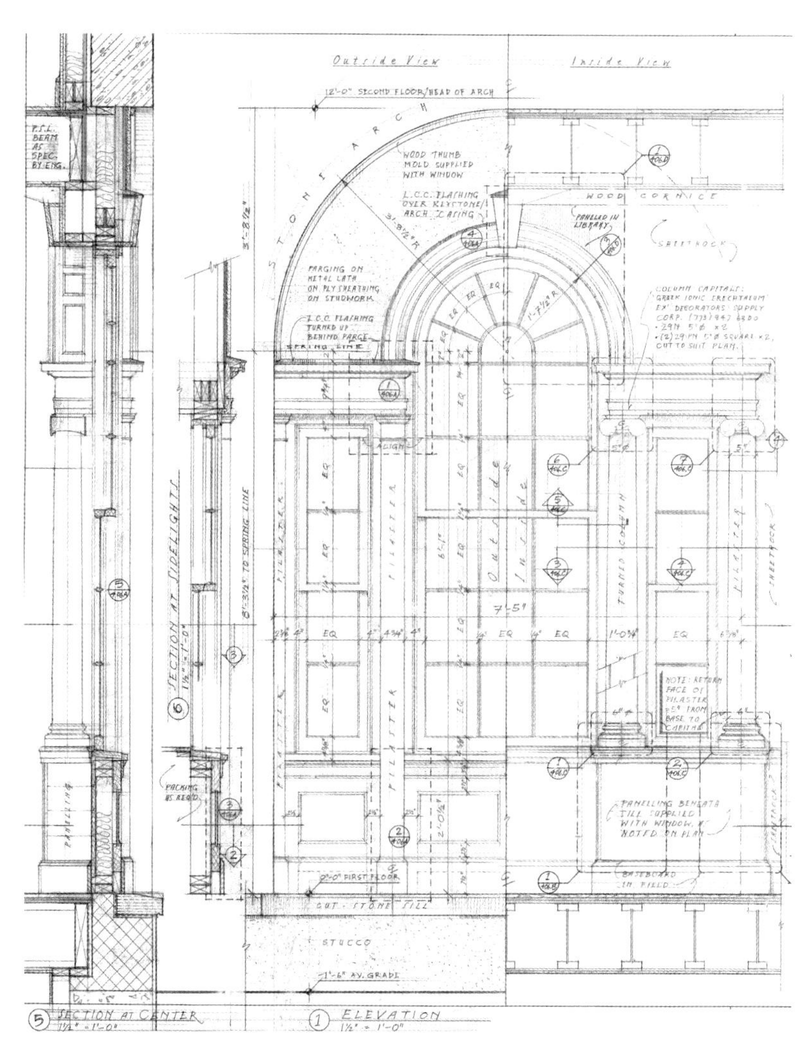
Outside View
Inside View
STONE ARCH
WOOD CORNICE
CUT STONE SILL
STUCCO
SECTION AT SIDELIGHTS
SECTION AT CENTER
ELEVATION

摄政风格建筑
Regency Style Estate

佛罗里达棕榈滩
Palm Beach, Florida

这座建于 20 世纪中期的房子，最初由棕榈滩著名建筑师 John Volk 设计。最近，Fairfax & Sammons 建筑设计公司对其进行了彻底翻修。

这次翻新为房子的新主人新增了客房、游泳馆和花园。房子的男主人是传媒大亨，妻子是记者，他们长年奔波于伦敦、纽约和多伦多，对夫妇二人来说棕榈滩是冬天的度假地。

这对忙碌的夫妇渴望能够忙里偷闲地在这里得到安静的休息。他们希望房子环境清新、舒畅——采光好，四面都能看到水景。这座房子得到了彻底的翻修，包括在起居室里采用新的建筑处理方法，起居室中还设有古典柱式顶部装饰，壁柱中间嵌入了建筑师定做的雕花壁炉架。

该住宅新增了一间图书室，并为夫妇两人各自设置了独立办公室，每间办公室都能观湖看海，建筑师还设计了一个两层的客房和一座游泳馆，并且对所有的客房进行了精装修——以满足两人众多来访朋友的要求。

为了能看到湖对面鸟类保护区的美景，建筑师新建了一个设有柱廊的阳台，这也避免了房子处于西晒之下。

这座典雅保守的摄政风格住宅，坐落在棕榈滩岛地产区的沃斯湖边。有直通大西洋私人海滩的地下隧道，这样的便利设施，在这个滨海豪宅林立的高档社区很是常见。

伦敦知名室内设计师米娜瑞克、亨利、泽维达其负责工程设计，弗吉尼亚州夏洛茨维尔市的查尔斯·斯迪克负责设计花园。

摄影：多斯顿·塞勒

This mid-20th century house was originally designed by the John Volk, a well-known Palm Beach architect in Palm Beach and was completely refurbished recently by Fairfax & Sammons Architects P.C..

A new guest house, pool house and garden design were incorporated into this renovation for its new owners, a media baron and his journalist wife based in London, New York and Toronto, for whom Palm Beach is a winter getaway.

Desiring a calm respite from their hectic lives, this busy couple wanted a fresh and airy ambiance – light-filled, with views to the water from every prospect. The house was given a complete make-over including a new architectural treatment in the living room with full entablature, engaged pilasters with a carved mantel commissioned by the architects.

A new library, two separate his and hers offices, each with their own view of lake and ocean respectively along with the two-story guest house and a pool pavilion were added and all guest rooms were given the luxe treatment – a requirement for their many visiting friends.

A new, fully colonnaded balcony was added to take advantage of the view of the spectacular bird sanctuary across the lake and to shade the house from the western exposure.

This elegantly restrained Regency styled house is sited in the estate section of the island of Palm Beach and perches on the edge of Lake Worth with an underground tunnel to a private beach on the Atlantic Ocean, a not-uncommon amenity for these opulent houses along the ocean in this exclusive community.

The celebrated London interior designers, Mlinaric, Henry, Zervudachi decorated the project, and Charles Stick of Charlottesville, Virginia designed the garden.

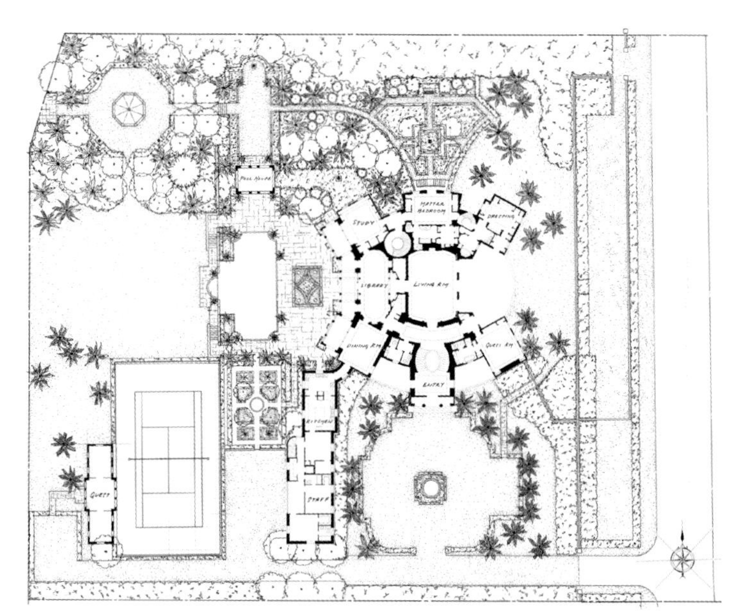
POOL HOUSE
MASTER BEDROOM
STUDY
DRESSING
LIBRARY
LIVING RM
DINING RM
GUEST RM
ENTRY
KITCHEN
STAFF

Granoff 建筑师事务所

Granoff Architects

Granoff 建筑师事务所创办于 1989 年，是一家提供全方位服务的公司，致力于提供杰出的客户服务和优秀的建筑作品。该公司拥有 22 名专业人员，工作室分别位于康涅狄格州的格林威治和纽约南安普顿市。他们从事的工程范围广泛，主要有四大类：住宅建筑、商业建筑、景观建筑和室内设计，其设计原则是在对过去的尊重和对未来的期盼之间做出平衡，从而设计出多样化、充满活力且品质持久的作品。

该事务所在不断地发展和成长。他们以独一无二的住宅、综合办公楼、多代住宅装修和壮观的景观设计而闻名。他们将自己的成功归功于与甲方紧密而长期的联系。他们对自己的作品充满活力与热情，要求自己的每一件作品都要超越上一件。他们运用想象力，从多角度看待问题，避免局限于单一的建筑风格。在 Granoff 建筑师事务所，建筑师们相信“绿色”设计是至关重要的，因此他们的每一项设计都包含了可持续性的策略。

他们应用广泛的工作经验、合理的公司经营理念和人力资源才能，以保证工程能够在预算内按时完成。

他们加强交流能力，以做到让人易于理解。建筑师们提出难题，用心聆听，然后为问题提供选择和专业见解。他们相信设计和建造的过程与最终的建筑作品一样重要，力图为甲方提供一次愉快的经历。

他们采用了一种与甲方共同设计的流程，寻找与他们的功能、经济和审美需求相一致的设计方案。他们的设计专业知识通过最先进的计算机辅助设计系统（CAD）来支持和加强。而且，他还有丰富的 3D 模型和计算机绘图经验。

Granoff Architects, established in 1989, is a full-service design firm dedicated to providing outstanding client service and achieving excellence in architecture. It has a professional staff of twenty-two, with offices located in Greenwich, CT and Southampton, NY. They work on a wide variety of projects in four integrated disciplines: residential architecture, commercial architecture, landscape architecture and interior design. Their design philosophy is balanced between respect for the past and excitement about the future, which has resulted in work of great variety, vitality and enduring quality.

Their practice has flourished and grown. They are well known for the design of unique homes, corporate offices, multi-family developments and spectacular landscapes. They credit their success to the close and long-term relationships they have with their clients. They conduct their work with energy and enthusiasm, challenging themselves to make each project even better than the last. They exercise their imagination, see problems from multiple perspectives and resist being limited to a single architectural style. At Granoff Architects, they believe that “Green” design is critical and incorporate sustainable strategies into each of their projects.

They apply their extensive professional experience, sound firm management philosophy and human resources capabilities to their projects to assure they are completed within budget and on schedule.

They are intensive, thorough communicators and remain accessible. They ask hard questions, listen thoroughly, then provide options and professional insight for answers. They believe that the process of designing and building is as important as the final built product, and strive to make the experience an enjoyable one for their clients.

They employ a collaborative design process with their clients and seek solutions which are consistent with their programmatic, economic and aesthetic needs. Their design expertise is supported and enhanced by our state-of-the-art computer-aided design (CAD) system. They have extensive 3D modeling and computer graphic capabilities.

作　品

Selected Work

中世纪住宅
Mid-Country Residence

康涅狄格州格林威治
Greenwich, Connecticut

这座约 740m^2 的住宅按照坚定的“绿色”理念来设计，将经典的石材和板岩设计结合起来。后屋顶上的光伏太阳能板每年可以产生 5kw 的电力。其他的绿色材料还包括非纤维玻璃保温材料、可再生的墙板、低挥发材料、竹地板和 FSC（森林管理委员会）认证的木材。该住宅和它的景观设计都完全符合其“富有挑战性”的地理位置。

This 740m^2 house combines a classic stone and slate design with a strong "Green" strategy. Photovoltaic panels at the rear roof will generate 5 kilowatts of electricity annually. Other green materials include non-fiberglass insulation, recycled wallboard, low VOC material, bamboo flooring and FSC certified lumber. The home and its landscape design fit comfortably onto the "challenging" site.

前立面

后立面

西侧立面

东侧立面

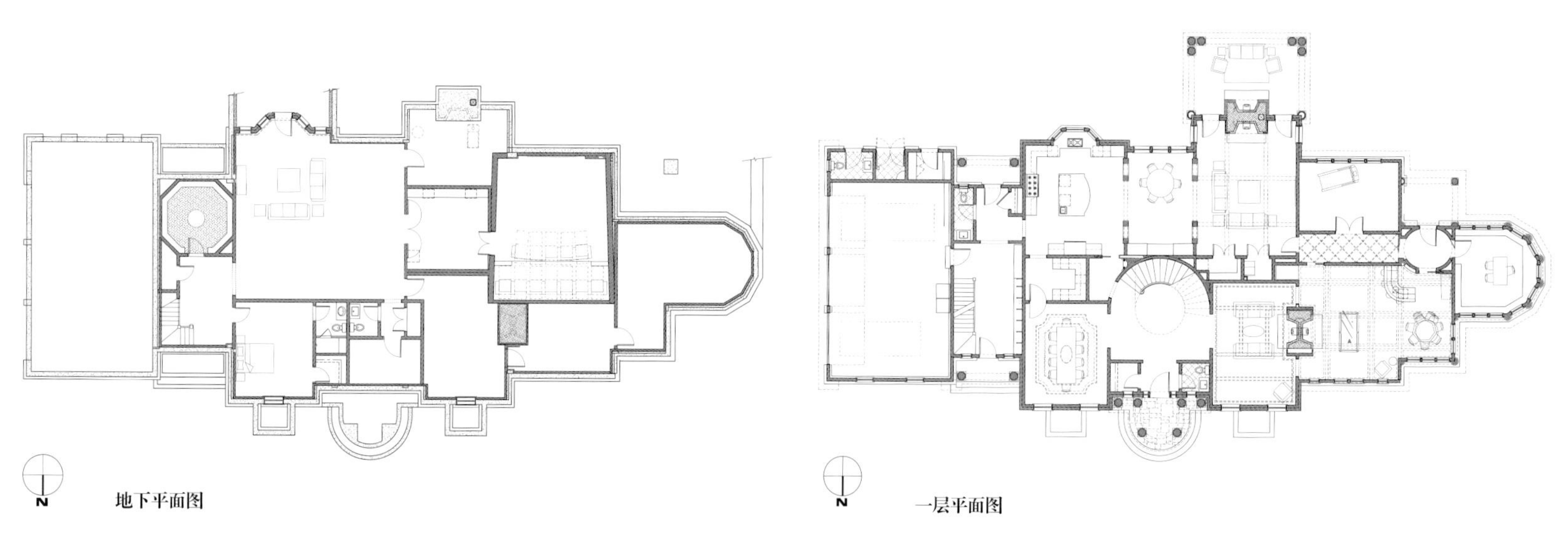
N
地下平面图
N
一层平面图

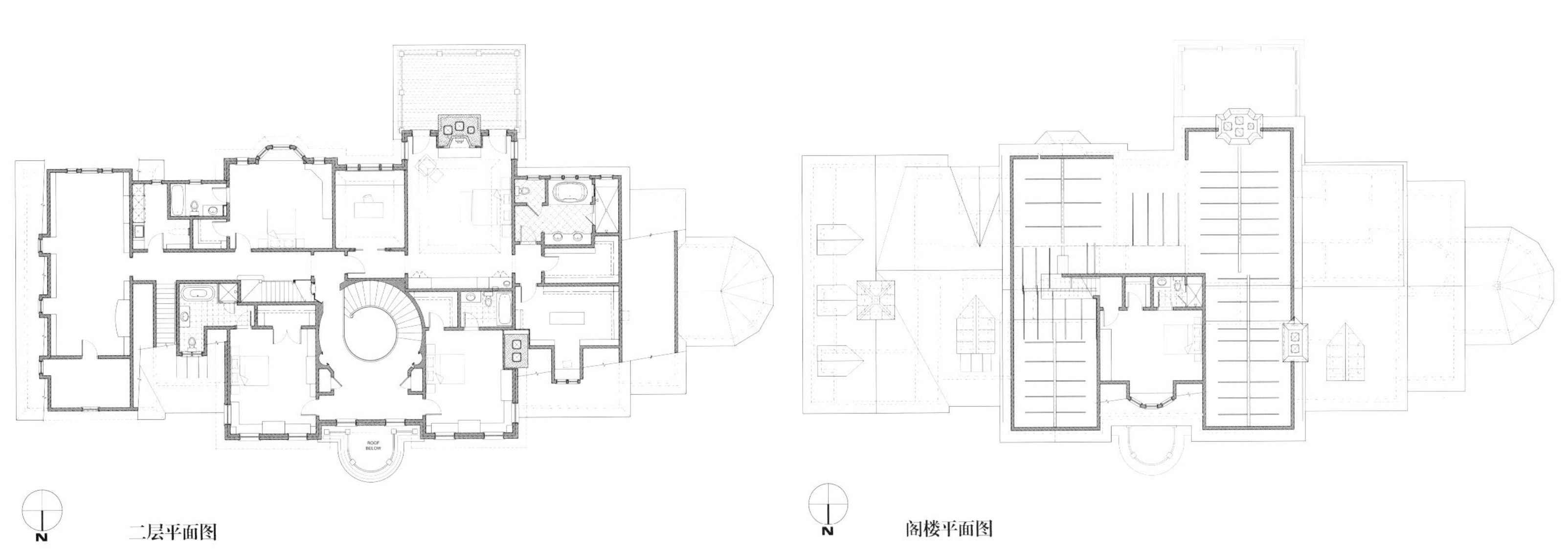
ROOF BELOW
N
二层平面图
N
阁楼平面图

私人住宅
Private Residence

纽约布里治汉普顿
Bridgehampton, New York

这座约 1300m^2 的建在约 68 796m^2 的令人印象深刻地块上的小山上，住宅为长条形的直角结构，大多数的房间都可远眺大海。如眉窗之类的传统的木板风格细部、塔楼、门廊和深悬壁结构都忠实于住宅的传统，而宏伟的入口大厅、双层高的起居室、品酒师的酒窖以及其他现代化设施则将住宅带回到了 21 世纪。然而真正让这处住宅与众不同的是其独一无二的细部设计，包括了生动的景观设计、带电梯的多层游泳池、再生石灰岩地板以及传统手工艺的灰泥作业。

Designed to sit on a knoll atop an impressive 1,300m^2 parcel, the long, angled structure of this 68,796m^2 home provides distant ocean views from most rooms. Traditional shingle-style details such as eyebrow windows, towers, porches, and deep overhangs are true to the home's classic roots, while a grand entry hall, two-story great room, sommelier's wine cellar and other modern amenities bring it solidly into the 21st century. But what really sets this home apart are the myriad one-of-a-kind details, from vivid landscaping and multi-tiered pools to an elevator, reclaimed limestone floors and artisanal plasterwork.

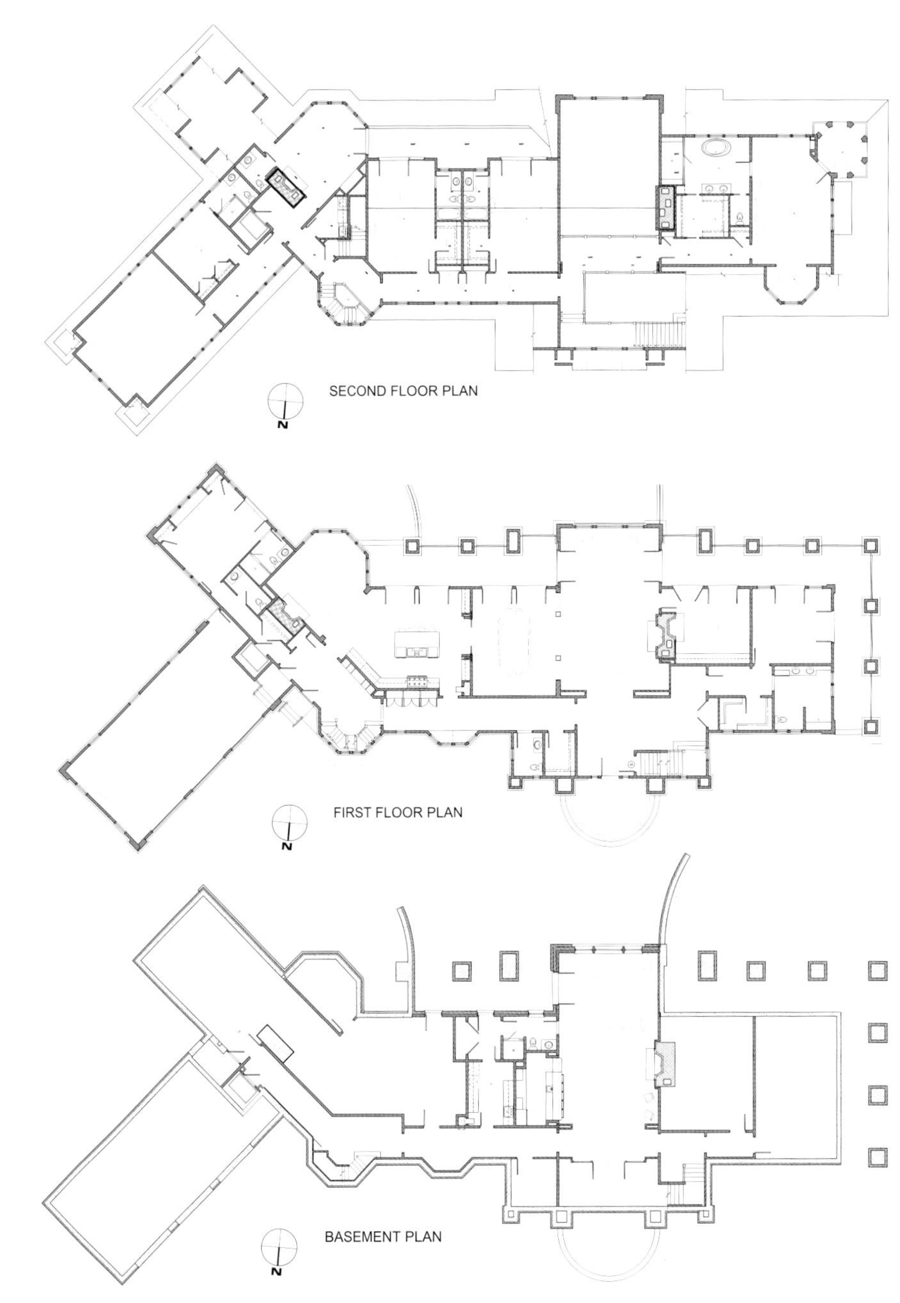
SECOND FLOOR PLAN
N
FIRST FLOOR PLAN
N
BASEMENT PLAN
N

SOUTH ELEVATION
WEST ELEVATION

NORTH ELEVATION

NORTHEAST ELEVATION

NORTHWEST ELEVATION

岩枫木住宅
Rock Maple

康涅狄格州格林威治
Greenwich, Connecticut

这座三层住宅的面积约840m^2，基地一侧为悬崖，另一侧则是一条河流。陡峭的倾斜屋顶轮廓线形成了强烈的视觉效果。受到世纪之交的木板风格住宅的启发，建筑师用敏锐的鉴别力对住宅的细部重新定义。车辆通道和石材的广泛应用使得住宅扎根在景观之中，而室内特色包括了三层高的楼梯间、超大尺寸的餐厅、樱桃木板装饰的书房以及有着教堂天花板的活动室。

Designed to stand comfortably between a massive rock out-cropping on one side and a river on the other, this 840m^2 home rises to a full three stories. The dramatic steep pitched roofline creates a strong curbside impression. Inspired by turn-of-the-century Shingle Style houses, the details are reinterpreted with flair. The porte cochere and extensive use of stone anchor the house into the landscape while the interior features a dynamic three-story stair hall, oversize dining room, cherry-paneled library and a playroom with a cathedral ceiling.

都铎风格住宅
Tudor Style Home

纽约威斯特郡
Westchester County, New York

这座砖石结构的都铎风格建筑结合了甲方的传统品位和建筑师的观点，占地约930m²。引人注目的板岩屋顶轮廓线、石拱和精细的石砌作业都使得住宅与景观完美融合。“蝴蝶”形的平面设计使得住宅内的许多房间都能够最大限度地享受阳光。通往住宅的迎宾小径和室内舒适的环境都说明这是一处受人欢迎的场所。

This brick and stone Tudor is a blend of the client's traditional tastes and the architect's vision for the site of this 930m² house. The striking slate roofline, stone arches and detailed masonry work help blend the house into the landscape. The "butterfly" plan maximizes the amount of daylight that fills the many dramatic spaces throughout the home. The inviting approach to the house and the cozy spaces within are a welcoming retreat.

SOUTH ELEVATION
WEST ELEVATION
EAST ELEVATION
NORTHWEST ELEVATION

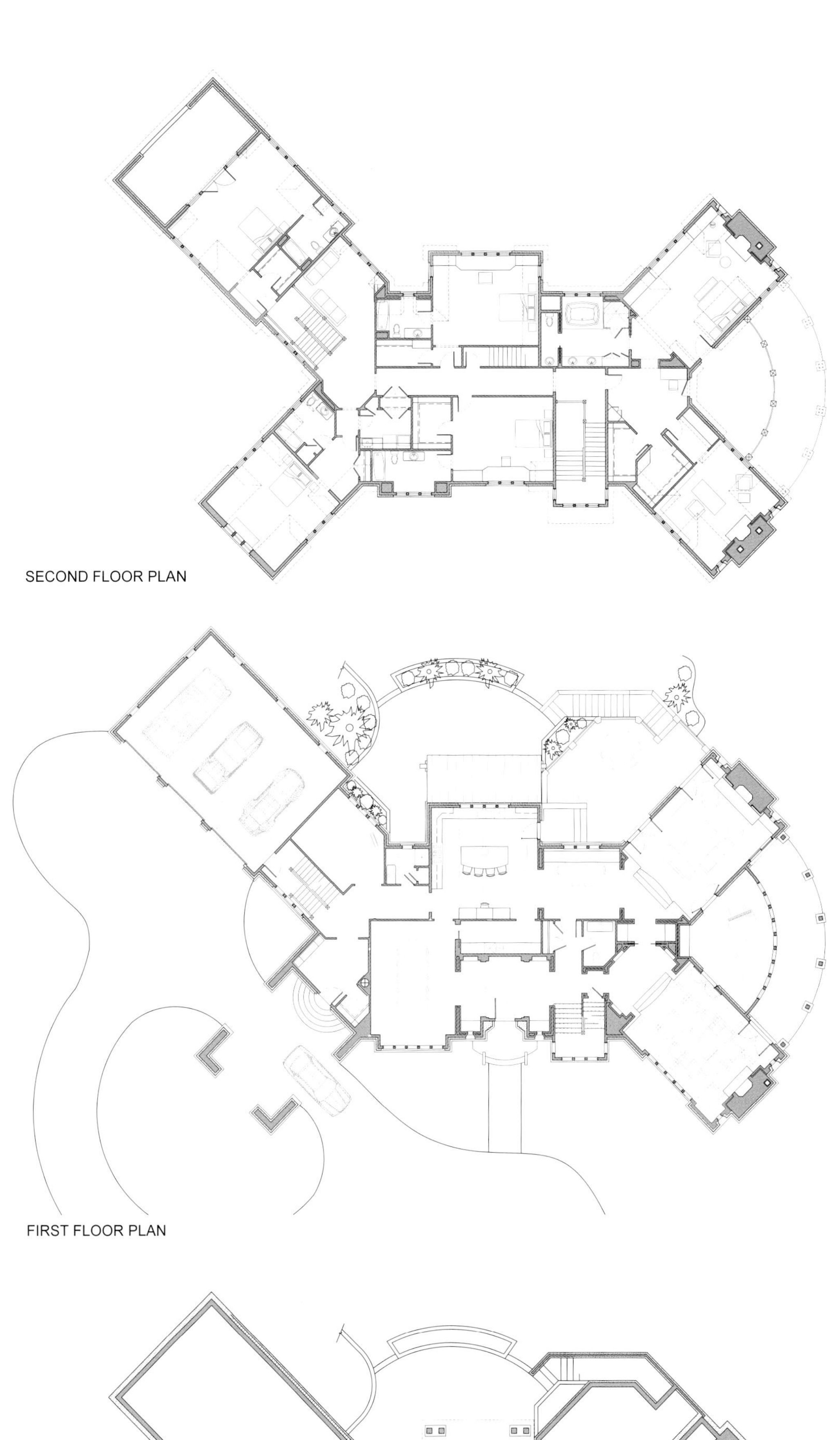

SECOND FLOOR PLAN

FIRST FLOOR PLAN

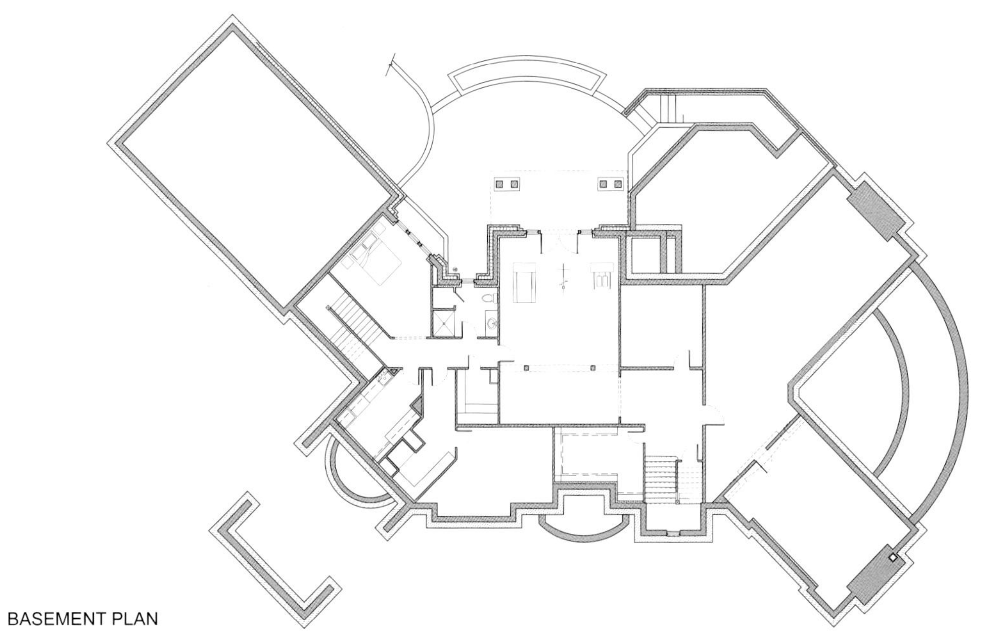

BASEMENT PLAN

兰德里设计集团

Landry Design Group

作为美国建筑师协会的成员以及兰德里设计集团的总裁，理查德·兰德里是世界上最负盛名的高端住宅建筑师之一。理查德·兰德里始终如一地为高品位用户提供尽善尽美的建筑，为他们筑起一个真正的家，其客户遍布 16 个国家，包括美国、加拿大、中国、日本和俄罗斯。

其公司兰德里设计集团成立于 1987 年，至今已为罗德·斯图尔特、舒格·雷·伦纳德、马克·沃尔伯格，汤姆·布拉迪和吉赛尔·邦臣设计过住宅。兰德里设计集团沉浸于建筑艺术当中，坚持奉行自己的美式折中主义品牌特色，借鉴最恰当的建筑风格，以满足客户独特的需求。

多年来，兰德里设计集团赢得了 2010 年美国赛区的"国际地产奖"、2008 年的"建造商选择奖"以及 47 项梦寐以求的"金块奖"。兰德里设计集团设计的住宅在过去四年里连续被《罗博报告》杂志评选为"终极住宅课题"。美国广播公司、家园频道、旅游频道、E! 娱乐以及 A&E 电视也都报道过该公司所设计的项目。

鉴于其在建筑领域的贡献，理查德·兰德里自 2000 年来就一直位列于《建筑文摘》的世界前 100 名建筑师和设计师的 AD100 名单中，并且自 2007 年以来在《罗博报告》杂志推荐的最佳建筑师名单中，他也一直榜上有名。2004 年，他被太平洋设计中心授予声名远播的"设计之星"的奖项。

兰德里设计集团已经在全世界完成了 400 多个项目，如今仍然致力于设计出能够迎合每个客户独特生活风格的住宅，与此同时，也展示了其在建造朴素优雅与含蓄唯美的住宅方面的无与伦比的能力。

Richard Landry, AIA, president of Landry Design Group, is one of the world's most celebrated high-end residential architects. With clients in 16 countries, including the United States, Canada, China, Japan and Russia, Richard Landry has consistently delivered impeccable architecture and a true home for those of discerning tastes.

Founded in 1987, his firm, Landry Design Group has designed homes for Rod Stewart, Sugar Ray Leonard, Mark Wahlberg, Tom Brady and Giselle Bundchen. Immersed in the art of architecture, Landry Design Group practices its own brand of American eclecticism, borrowing the most appropriate architectural styles to answer its clients' unique needs.

Over the years, Landry Design Group has received the 2010 International Property Award – Americas Division, the 2008 Builder's Choice Award and 47 coveted Gold Nugget Awards. Homes designed by Landry Design Group have also been selected for *Robb Report*'s "Ultimate Home issue" for the last four consecutive years. The firm's projects have also been featured on ABC, HGTV, Travel Channel, E! Entertainment and A&E Television.

In recognition of his contributions to the field of architecture, Richard Landry has consistently been cited on *Architectural Digest*'s AD100 list of top 100 architects and designers since 2000 and has been included on "Robb Report Recommended" list of best architects since 2007. In 2004, he was presented with the prestigious "Stars of Design" award from the Pacific Design Center.

With a portfolio of over 400 projects around the world, Landry Design Group continues to design residences sensitive to the unique lifestyles of each client while demonstrating the firm's unparalleled skill in creating residences of understated elegance and beauty.

作　品

Selected Work

富兰克林住宅
Franklin Residence

Haselton 住宅
Haselton Residence

富兰克林住宅
Franklin Residence

加利福尼亚州
California

安达卢西亚人的乐园

兰德里设计集团打造了一栋高雅住宅，该栋住宅深受西班牙风格影响，周围环绕着植根于自然的热带乐园，人们在这里可以俯瞰到城市的欢乐景象。

这栋住宅为之前的建造者带来了不可避免的挑战。建造地点位于一片沿山坡延伸的带状区域。住宅的入口位于车行道的末端，而车行道又正好位于该基地中最狭窄的部分。为了最大限度地保护这片土地，同时也能让人们在到达这片区域时就可以一览无余地欣赏到该住宅的景观，理查德•兰德里设计了一家带有地下停车场的石质汽车旅馆。

为了保持一些原有的印迹，兰德里和他的团队重新设计了楼层平面图，以使景观和私密性达到最大化。同时为了方便出入以及形成视觉冲击，建筑师也将房间进行了重新定位。公共空间和正式的花园现在位于住宅的前部，更私密和非正式的空间则隐匿在住宅的更深处。

建筑师将原有结构拆除得干干净净。之前大部分房间都缺少充足的光线，并且完全忽略了城市景观，现在这些房间都向室外开放。建筑师之所以选择现代化的安达卢西亚建筑风格，是因为它能够容纳宽敞的窗户以及消失不见的门，这让人们能够在地中海般的气候中舒适地生活。

同时，采用这样的建筑风格是因为雕刻的木材和安达卢西亚住宅中的简约结构与环太平洋地区建筑的细节处理方法和吸引业主的水元素融合在一起。两个水母缸位于主卧室壁炉的两侧，两个热带水箱是客厅的特色所在。所有之中最引人注目的是一个5m长的鲨鱼缸，它将客厅和餐厅分隔开来。

建筑师团队与Franco Vecchio工作室通力合作，将海洋那充满活力的色彩运用到了墙面处理过程和舒适的家具中来。兰德里公司也就精心设计的度假胜地般的水上游乐场征询了他人意见。9m长的泳道延伸到山坡中的挖空部分，它将两个大型水池分离开来。锦鲤池、水上滑梯、温泉、游泳蒸汽洞穴和池畔酒吧使水池熠熠生辉。

为了能够充分利用这些壮丽景色，主要的水景贯穿在整栋住宅中。客厅、家庭娱乐室、餐厅和厨房都朝向有屋顶的大型长廊。在其远处是一个设有大型喷泉的露台，水可以涌进下方草地旁边的水池中。喷泉中的防火元素安装在睡莲的后面，是住宅中另一个以水为主题的实例。

客厅通过一条私密的长廊可以通向瀑布和一条蜿蜒的“河流”。毗邻水池的休息区拥有以景观为主的屋顶休息区域，周围环绕着具有浪漫风情的壁炉。这栋令人叹为观止的住宅及其具有异域风情的景观反映出理查德•兰德里在设计独一无二的建筑时所运用的方法。

摄影：Erhard Pfeiffer

An Andalusian Paradise

Landry Design Group created an elegant Spanish-influenced estate surrounded by a tropical paradise grounded in nature and overlooking a city's delights.

This home site presented inescapable challenges to previous builders. It is a long ribbon of land that runs along a hillside. The entry to the house is at the end of a drive on the narrowest part of the property. To preserve as much of the land as possible and to maintain an unobstructed view of the house upon arrival, Richard Landry designed a stone motor court with underground parking.

Maintaining some of the original footprint, Landry and his team reconfigured the floor plan to maximize views and privacy. Rooms were also relocated for convenience and visual impact. The public rooms and formal gardens are now in the front part of the residence. The more intimate and casual spaces are located deeper into the property.

The original structure was stripped down to the studs. Most of the rooms, previously lacking adequate light and completely ignoring city views, now open to the outdoors. An updated Andalusian style was selected because it would allow for expansive windows and disappearing doors that could breathe in the Mediterranean-like climate.

The architectural style was also adopted because the carved wood and rustic components associated with homes in Andalusia blend with the Pacific Rim details and water elements that appeal to the owner. Two jellyfish tanks flank the fireplace in the master bedroom and two tropical tanks are featured in the great room. Most dramatic of all is a 5m shark tank that separates the living room from the dining room.

The architectural team worked with Franco Vecchio Studio to translate the vibrant colors of the ocean into wall treatments and relaxed furnishings. Landry's firm also consulted on the elaborate, resort-like water playground. A 9m swim channel extends into a hollowed-out part of the hillside and separates two large pools. Energizing the waterways are a koi pond, water slide, spas, a swim-in steam cave and a swim-up bar.

To make the most of the spectacular setting, seating with prime water views is arranged throughout the property. The living room, family room, dining room and kitchen open to a large covered loggia. Beyond it is a patio with an infinity-edge fountain that spills into a basin near the lawn below. Fire elements in the fountain are modeled after water lilies, another example of the estate's water leitmotif.

The great room looks through an intimate loggia toward waterfalls and a wandering "river." A lounge area adjacent to a pool has view-oriented rooftop seating around a romantic fireplace. This stunning house and its exotic landscape reflect Richard Landry's approach to create unique estates that become the ultimate escape.

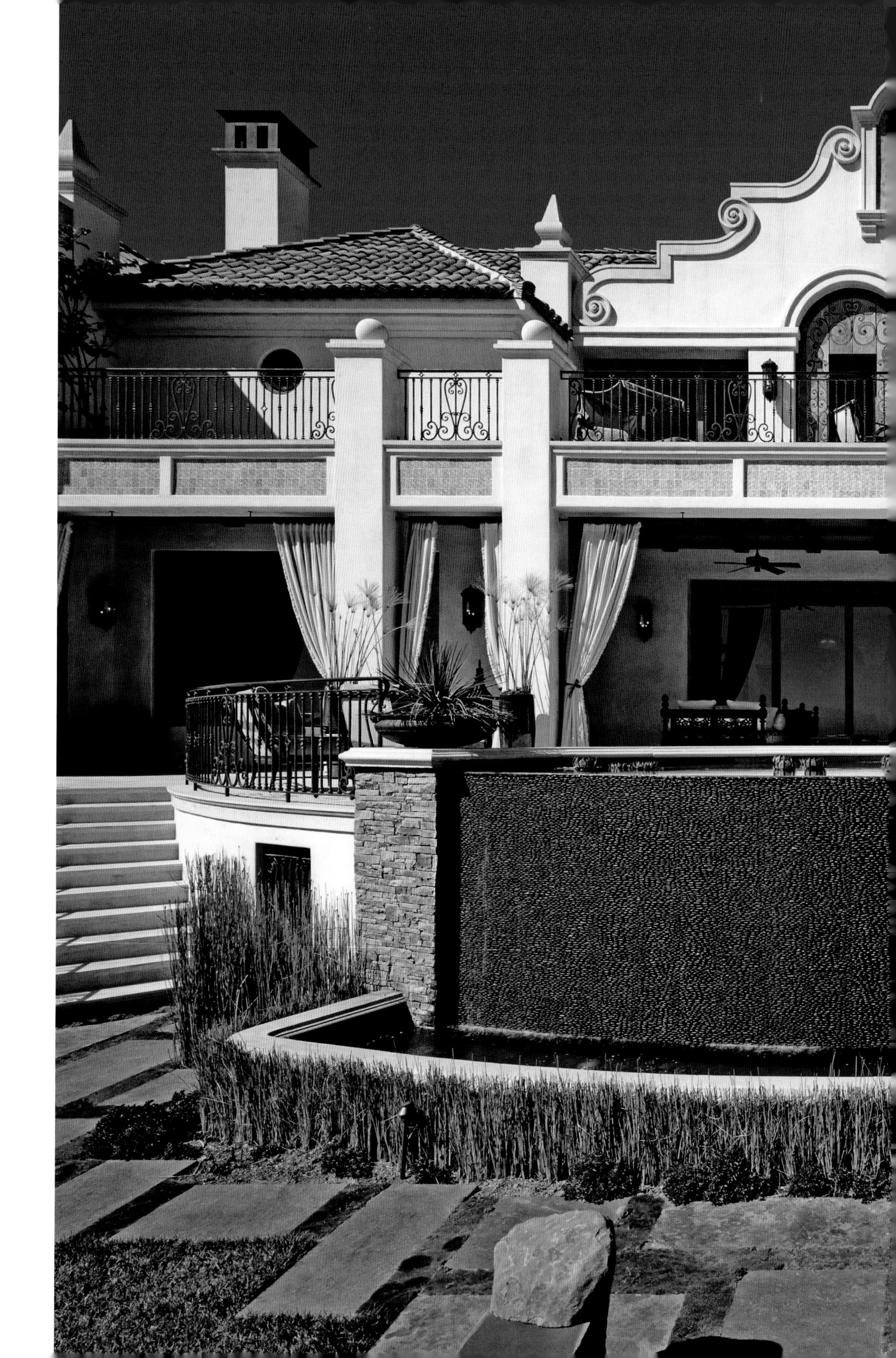

ELVIS

Haselton 住宅
Haselton Residence

加利福尼亚州
California

兰德里设计集团在一开始设计 Haselton 住宅时便面临着许多内在限制的挑战。这座位于南加利福尼亚州的小镇，它引以为傲的历史文化让人赞叹不已，这片土地过去曾是西班牙殖民地，这一点也在当地的地中海建筑风格中体现得淋漓尽致。它的建筑审查委员会审查了所有新建建筑的建造计划。因此，最终的设计方案不仅要满足客户的需求，同时也要适合小镇的环境。

在各种限制条件下，兰德里打造了一栋低调的地中海风格别墅，宛如一座西班牙小山村：大部分的建筑都是一层的，屋顶的轮廓线上覆有火红色的黏土瓦，变化多样，它不断延伸直至住宅的侧翼建筑，这部分建筑上升到了二层的位置。建筑外部是手工用泥刀涂抹的光滑灰泥，灰泥有着温暖的赭石色调，它是一栋在汽车庭院中心与巨大的橄榄树形成互补的美丽建筑。

该项目坐落在山坡上，人们在 Haselton 住宅的每一个房间内都能够俯瞰到周围山脉和海湾的秀美景色。同时建造地点非常私密，周围环绕着起伏的群山，成熟的景观使其可以藏匿于附近的住宅中。

室内的楼层简洁且流畅。SFA 设计公司负责室内设计，该公司将客户所使用的大部分房间都连接起来。客厅设有 8m 高的桁架天花板、一个石灰岩壁炉以及由托斯卡纳支柱支撑的拱门。拱形的法式大门通向室外有屋顶的门廊。由拱门构造成的用餐区是客厅的一部分，厨房设有经过专门设计的桤木橱柜，也与客厅连在一起。

Haselton 住宅的天花板是其最鲜明的特点。这是兰德里设计集团的特有手法，他们为这片常常被忽略的区域注入了众多的创意理念，也投入了大量的精力。在这个项目中，每个房间的天花板都与众不同。有拱形的、梁的、桁架式的，或是有壁画装饰的，每一种都各具特色。

在地下室的部分区域中，酒窖位于定制锻造铁门的后面，这是一个面积不大但非常别致的房间。这里有装饰着仿古砖的筒形天花板，以及用来存放稀世葡萄酒的通高雪松货架。天花板和架子之间的模型是用手工雕刻的，上面雕刻着关于葡萄藤和葡萄串的葡萄园图案。在后面的墙上，一幅具有独创性的 trompe–l'oeil 壁画生动地描绘出了 Haselton 住宅的酒窖的景象，酒窖里堆满了木质葡萄酒酒桶。

客厅、厨房和用餐区都朝向独特的半月形门廊敞开，在这里可以俯瞰到倒映池和喷泉。这里设有一个壁炉，其周围镶嵌着定制的陶瓷瓷砖。在这里，人们可以欣赏到远处群山和海洋的壮丽景色，因此这里就成了一处为非正式娱乐活动打造的完美空间。

图书馆在通往主卧室的途中。这里的住户都是忠实的读者，所以兰德里设计了一个引人注目的房间，人们可以在这里流连徘徊：通高的书架，颜色深沉、年代久远的桃花心木镶板，以及可以驱走海边寒意的大理石壁炉。在天花板的中心部分，金箔圆孔支撑着黄铜枝形吊灯。

主卧室在细节设计方面也非常丰富，这里有高高的圆形平顶窗户以及向内深深凹陷的天花板，这些都着重强调了意大利式的壁画风格。嵌入式的壁龛形成了西班牙风格的床的框架，也运用了手绘细节处理技术。由西班牙塞维利亚的艺术家绘制的壁画设置在陶瓷瓷砖上，描绘了威尼斯大运河和圣马可的壮丽景色，人们可以通过大窗户看见这幅画。

Haselton 住宅受西班牙殖民地这一传统影响很大，理查德·兰德里的设计也借鉴了不同时代和不同地区的设计理念，并将这些融入到他的世俗眼光中来。这种建筑风格扎根于建筑传统之中，但却并没有被它束缚。

摄影：Erhard Pfeiffer

Landry Design Group's design for the Haselton home had a number of built-in restrictions from the very beginning. This small town in Southern California, is justifiably proud of its historic, Spanish Colonial past, clearly reflected in its Mediterranean architecture. Their architectural review board would review all new construction plans. Thus the final design had to be appropriate not just for the clients, but for the town, as well.

Within their strictures, Landry produced a low-key Mediterranean villa resembling a small, Spanish hillside village: mostly one story, the roofline topped with burnt-red clay tiles, varies and rambles until it reaches the one wing of the house that rises to the second story. The exterior is hand-troweled smooth stucco in a warm, ochre color, a lovely complement to the huge olive tree in the center of the automobile court.

Situated on a hillside, each room of the Haselton house has a commanding view of the surrounding mountains and bay. The location is intensely private; surrounded by rolling hills and concealed from neighboring homes by mature landscaping.

The inside floor plan is simple and fluid. Appointed by the interior design firm of SFA Design, all of the rooms used most by the clients are connected. The great room features an 8-meter-high trussed ceiling, a limestone fireplace and arches supported by Tuscan pillars. Arched French doors lead to the covered loggia outside. The dining area is part of the great room, framed by the arches, and the kitchen, with its custom-designed alder wood cabinetry, is connected to it as well.

The ceilings of the Haselton house are among its most distinctive features. It is a signature of the Landry Design Group to devote a great deal of their creative attention and energy to this often-neglected area, and in this project, each room's ceiling is very different. Coved, beamed, trussed, or frescoed, each has its own character.

In the partial basement, behind a custom-designed, wrought iron door, is the wine cellar, a small, but extraordinary room. It features a barrel ceiling of antique bricks, and floor-to-ceiling cedar racks for storing rare vintages. The moldings between the ceiling and the shelves are hand carved with a vineyard motif of vines and grape clusters. On the rear wall, a clever trompe-l'oeil mural depicts a mirror image of the Haselton's wine cellar, filled with wooden wine casks.

The living room, kitchen and dining area all open up to the unusual, demi-lune shaped loggia, overlooking a reflecting pool and fountain. It features a fireplace surrounded by custom ceramic tile. It offers a breathtaking view of the hills and the ocean beyond, a perfect space for informal entertaining.

The library is en route to the master bedroom. The clients are avid readers, so Landry designed a seductive room that invites one to linger: floor-to-ceiling bookshelves, dark, mellow paneling all in mahogany, and a marble fireplace to chase away the coastal chill. In the center of the ceiling, a gold-leaf oculus supports a brass chandelier.

The master bedroom is richly detailed with tall, round-topped windows and a deeply coved ceiling, highlighted by an Italian-style fresco. A recessed alcove framing the Spanish bed also features hand-painted detailing. A mural, painted by an artist in Seville, Spain, on ceramic tile, depicting the Grand Canal of Venice and San Marco, can be seen through a large window.

While the Haselton house, is heavily influenced by the Spanish Colonial tradition, Richard Landry's design also borrows from different eras and regions, integrating them into his worldly vision. It is anchored in an architectural tradition without being shackled by it.

Oatman 建筑设计公司

Oatman Architects, Inc.

美国建筑师协会的 Homer Oatman 是 Oatman 建筑设计公司的总裁和首席建筑师。Oatman 建筑设计公司坐落于加利福尼亚州的纽波特比奇市，专门从事整个美国西部、拉丁美洲和中东的豪华住宅项目和精品度假酒店的设计。

Homer Oatman 本科就读于斯坦福大学和加利福尼亚大学的圣迭戈分校，取得艺术学士学位，并取得加利福尼亚大学洛杉矶分校建筑设计与城市规划学院的建筑学硕士学位。

在创立 Oatman 建筑设计公司之前，Homer Oatman 曾经是 KTGY 的合作伙伴和首席设计师，并且担任位于加利福尼亚州纽波特比奇市的尔湾公司的定制住宅的设计总监。

Homer Oatman, a member of AIA is the president and principal designer of Oatman Architects, Inc. located in Newport Beach, California, specializing in the design of luxury residential projects and boutique resort hotels throughout the Western United States, Latin America, and the Middle East.

Homer Oatman received his undergraduate education at Stanford University and the University of California at San Diego with a bachelors degree in Fine Arts. He received his master degree of Architecture from the Graduate School of Architecture and Urban Planning at UCLA.

Prior to founding Oatman Architects, Inc. Homer Oatman was a partner and design principal at KTGY and director of Design for Custom Homes for The Irvine Company in Newport Beach, California.

作　品

Selected Work

水晶湾意大利北部风格住宅 1（阿尔伯特住宅）
Crystal Cove Northern Italian Style Residence 1 (The Albert Residence)

水晶湾意大利北部风格住宅 2（Oates 住宅）
Crystal Cove Northern Italian Style Residence 2 (The Oates Residence)

水晶湾安达卢西亚风格住宅（白昼住宅）
Crystal Cove Andalusian Style Residence (The Day Residence)

约山圣巴巴拉式住宅（Hahn 住宅）
Covenant Hills Santa Barbara Style Residence (The Hahn Residence)

水晶湾意大利北部风格住宅 1（阿尔伯特住宅）
Crystal Cove Northern Italian Style Residence 1 (The Albert Residence)

加利福尼亚水晶湾
Crystal Cove, California

要在纵深地块建造建筑面临的挑战是要将场地前面一直到主要空间所在的后面连接起来。这将通过设计一条开放性走廊来实现，这条走廊穿过传统风格的房屋正面体量，并且开口正对入口庭院，可以看到部分海景。参观者将由一条环绕庭院的带顶凉廊引导来到位于住宅平面图中心的真正正门处。每个最初进入大厅的人都会被那奇特的三层开放性环形楼梯所吸引。当参观者站在大厅中央时，他将会由于看到沿着住宅的中轴线壮观的景色而将注意力向左转90°，视线穿过中轴线和后面带柱的凉廊，越过一条没有边际的池塘而终止。这个池塘看起来一直流入低于地平线一百多米的太平洋。

The challenge with a deep view lot is to provide an engaging procession from the front of the lot to the rear where the primary spaces are located. This was achieved by designing an open breezeway piercing the traditional frontal massing of the house and opening onto an entry courtyard with a partial view of the ocean. The visitor is led under a covered loggia around the courtyard to the actual front door which is in the center of the house plan. Initially entering the foyer one's view is focused on a dramatic three-story open circular stair. Standing at the center of the foyer, the visitor's focus is turned 90-degree left by a dramatic view through the central axis of the house terminating through the rear loggia colonnade, over an infinite edge pool which seemingly spills into the Pacific Ocean nearly over 100 meters below uninterrupted to the horizon.

SECOND FLOOR PLAN

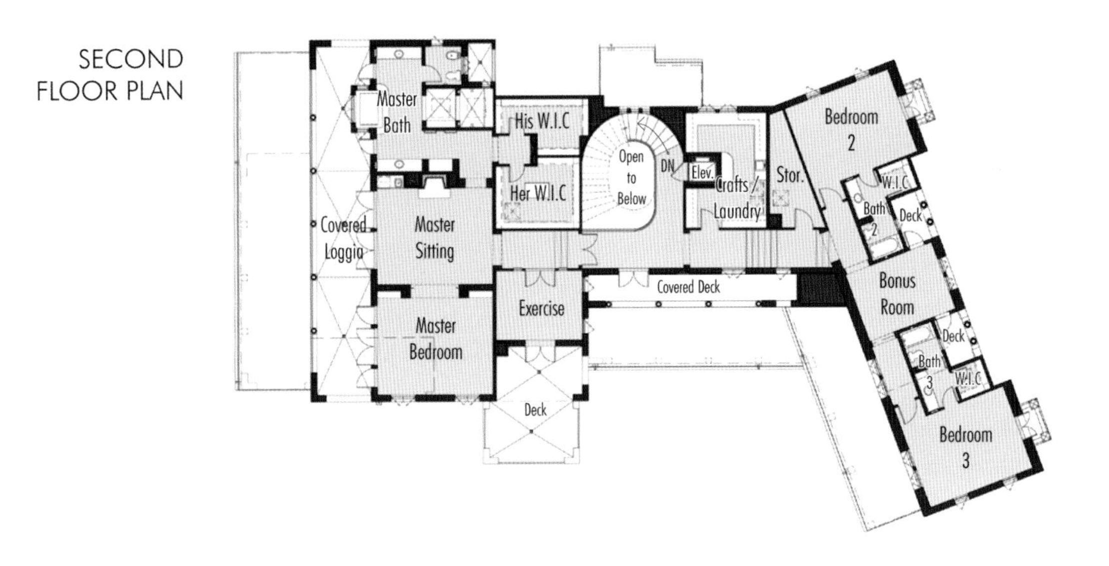

FIRST FLOOR PLAN

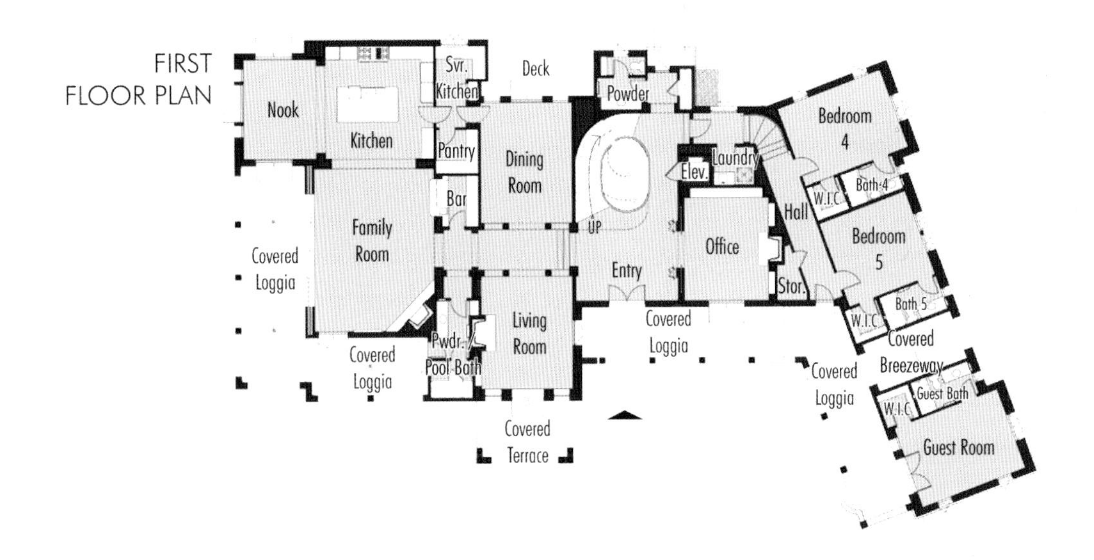

BASEMENT FLOOR PLAN

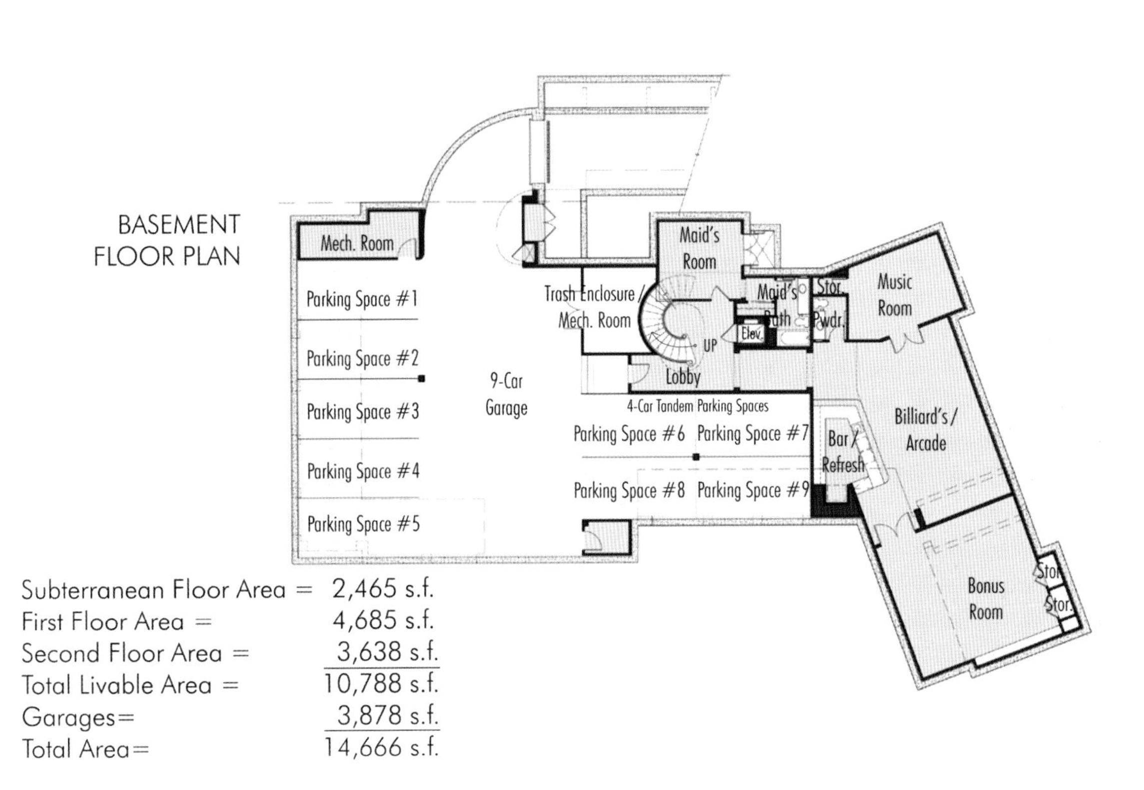

Subterranean Floor Area =	2,465 s.f.
First Floor Area =	4,685 s.f.
Second Floor Area =	3,638 s.f.
Total Livable Area =	10,788 s.f.
Garages=	3,878 s.f.
Total Area=	14,666 s.f.

LOOKING T'WARDS LIVING ROOM

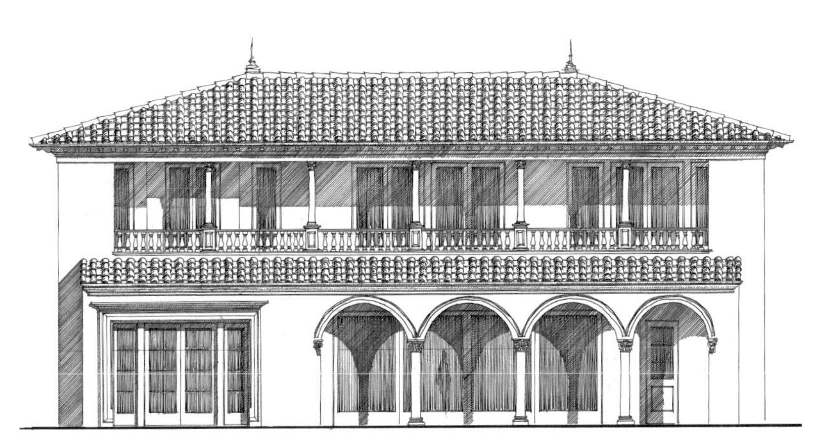

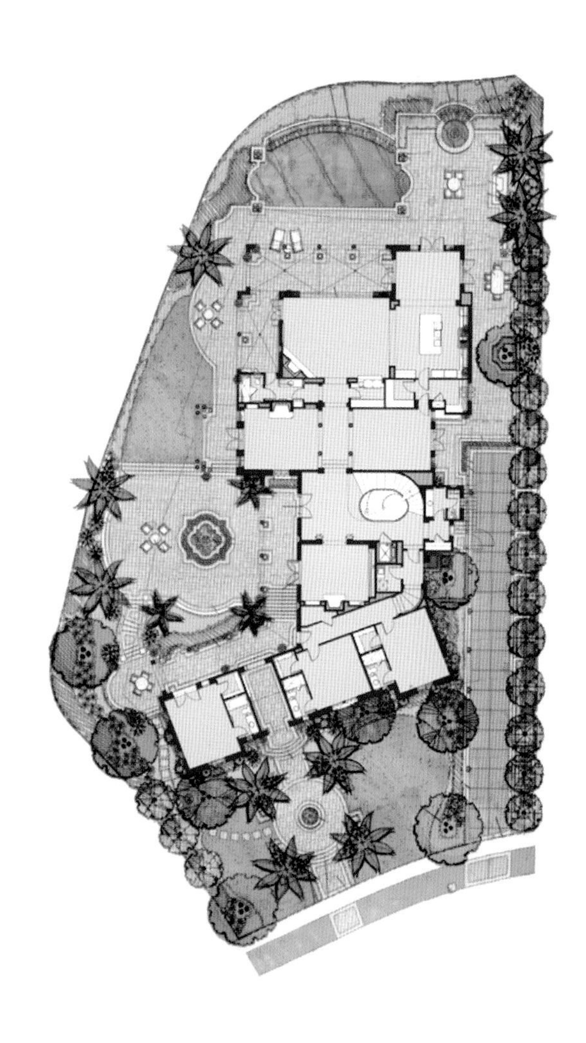

水晶湾意大利北部风格住宅 2（Oates 住宅）
Crystal Cove Northern Italian Style Residence 2 (The Oates Residence)

加利福尼亚州水晶湾
Crystal Cove, California

这是一块狭长纵深的地块，从东（前面）到西（后面）延伸，主要景色集中在地块后面。建筑的平面布局呈 L 形，较短的边垂直于前面的建筑红线以便于使街道的影响最大化，较长的边平行于北面的建筑红线，这条长边打开了该地块的南部，为一个长条形健身游泳池提供了空间。游泳池所在的地点在建筑阴影以外，常年受阳光照射，尤其是冬季。

This was a narrow deep lot running east (front) to west (rear) with a primary view to the rear. The floor plan was generally configured in an "L" shape with the short leg perpendicular to the front property line to maximize street impact .The long leg runs along and parallel to the north property line. This opened up the southern length of the lot to provide space for a long lap pool which has year-round sun outside of the building shadow, particularly in winter.

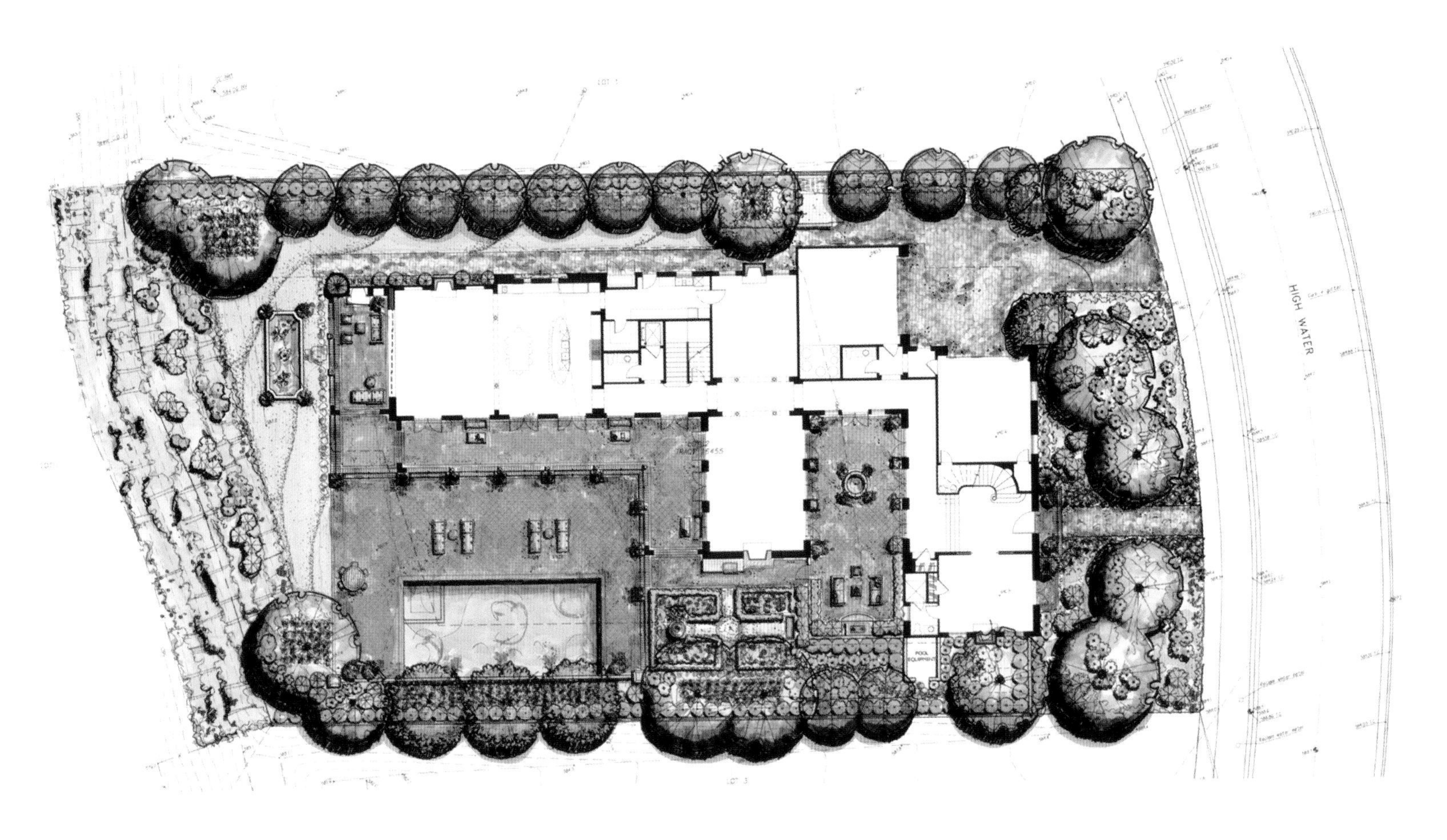
HIGH WATER

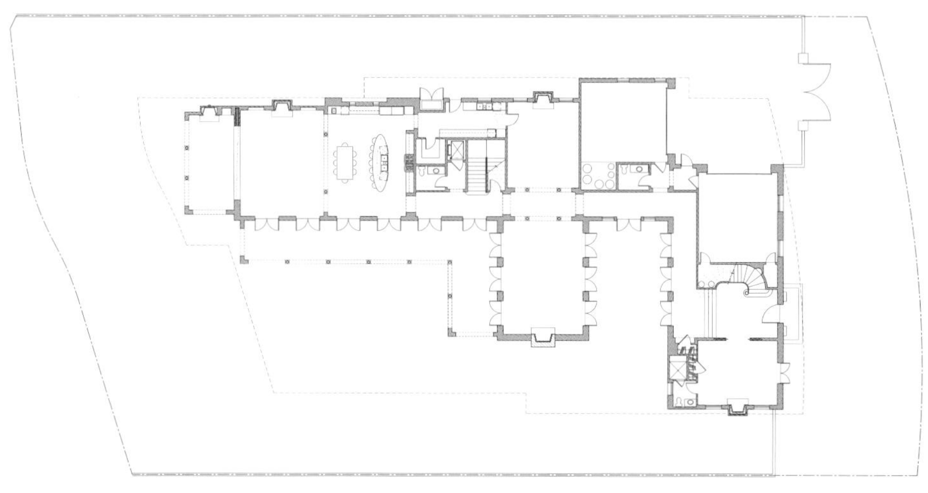

FEBRUARY 06, 2007

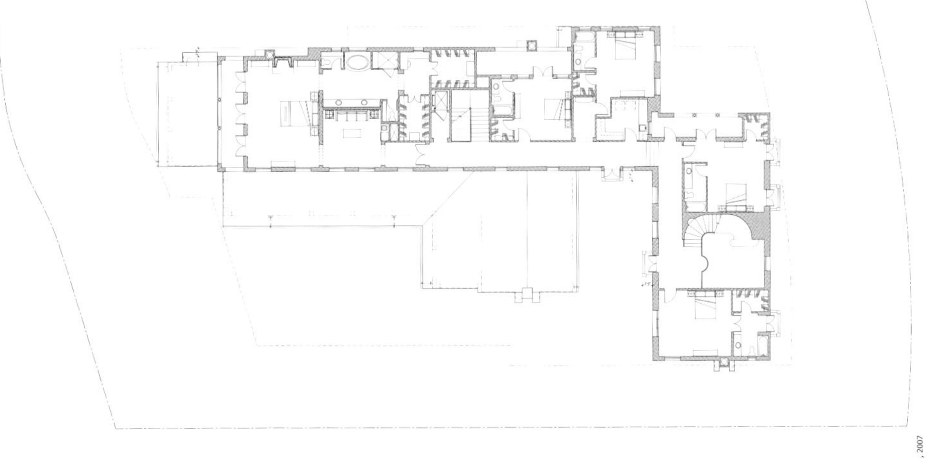

FEBRUARY 06, 2007

水晶湾安达卢西亚风格住宅（白昼住宅）
Crystal Cove Andalusian Style Residence (The Day Residence)

加利福尼亚州水晶湾
Crystal Cove, California

这栋住宅是围绕着一个与地下室齐平的中央庭院设计的，水平庭院为地下室的娱乐活动室和健身房提供光线、空气和室外活动空间。该住宅的入口周围是按照 19 世纪西班牙塞维利亚住宅的细节设计的。

This house is designed around a central basement level courtyard that provides light, air and an exterior social space to the basement entertainment room and gymnasium. The entry surrounding was designed from details of historic 19th century residences in Sevilla, Spain.

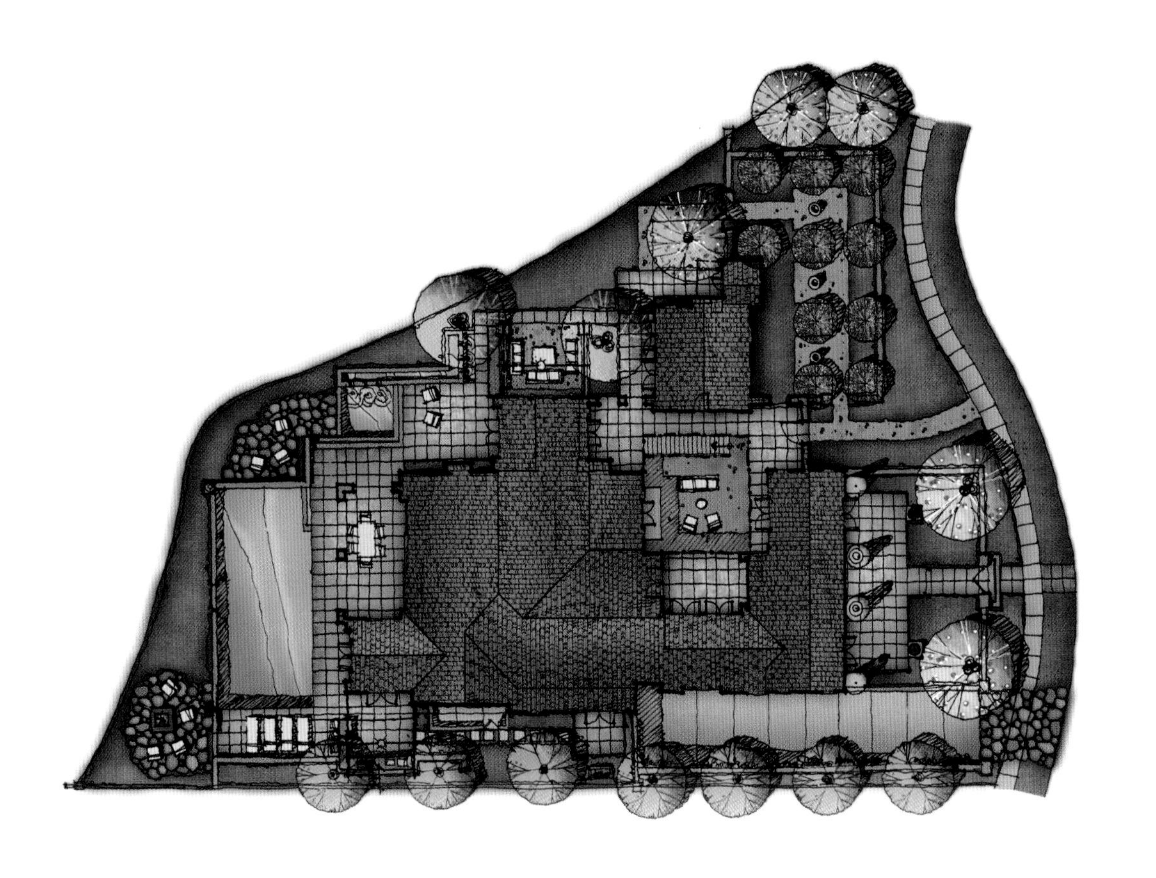

FULL SCALE MOCK UP
TO BE PROVIDED
DURING TOOLING
FULL SCALE MOCK UP
TO BE PROVIDED
DURING TOOLING
ENTRY TO ZAGUAN
SCALE: 3/8" = 1'-0"

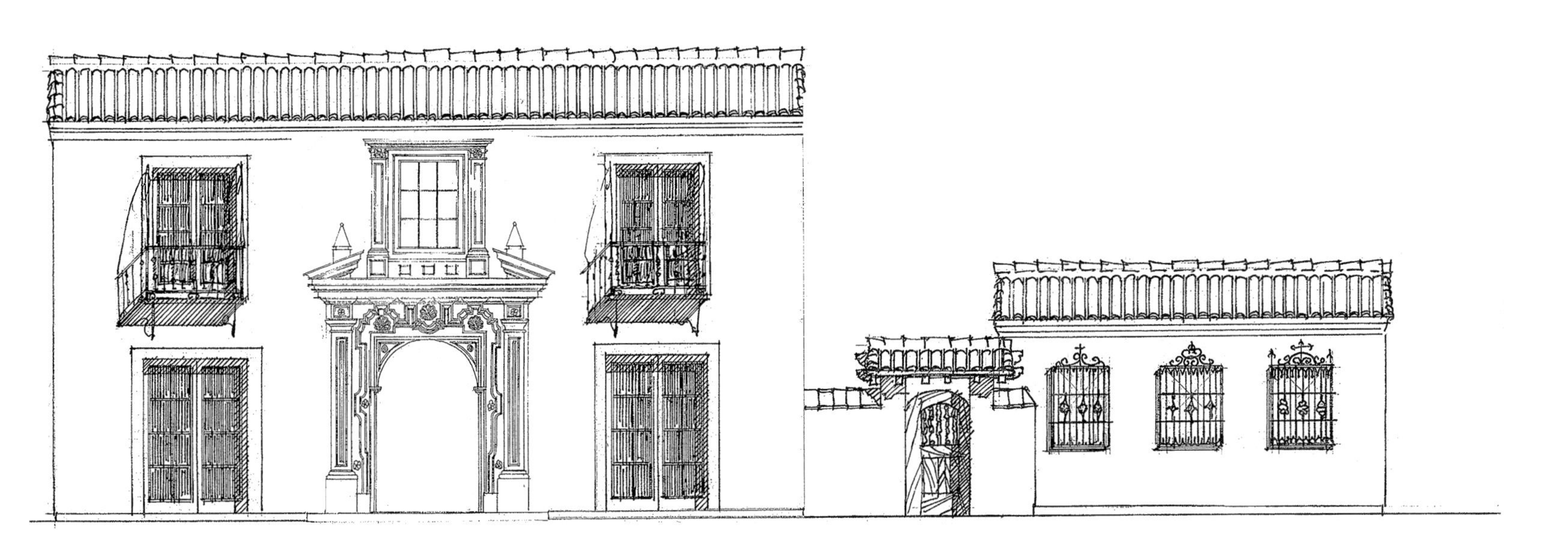

约山圣巴巴拉式住宅（Hahn 住宅）
Covenant Hills Santa Barbara Style Residence (The Hahn Residence)

加利福尼亚州，Ladera 农场，约山
Covenant Hills, Ladera Ranch, California

这栋住宅是按照 20 世纪 20 年代文艺复兴时期圣巴巴拉式休闲方式设计的。为了达到那个时期细节上的逼真性，整栋住宅都采用西班牙定制的手工瓷砖和当地的墨西哥锻铁。

This house is designed in the playful manner of Santa Barbara style Mission Revival houses of the 1920's. Custom handmade tiles from Spain were used throughout the house along with local Mexican wrought iron for authentic period detailing.

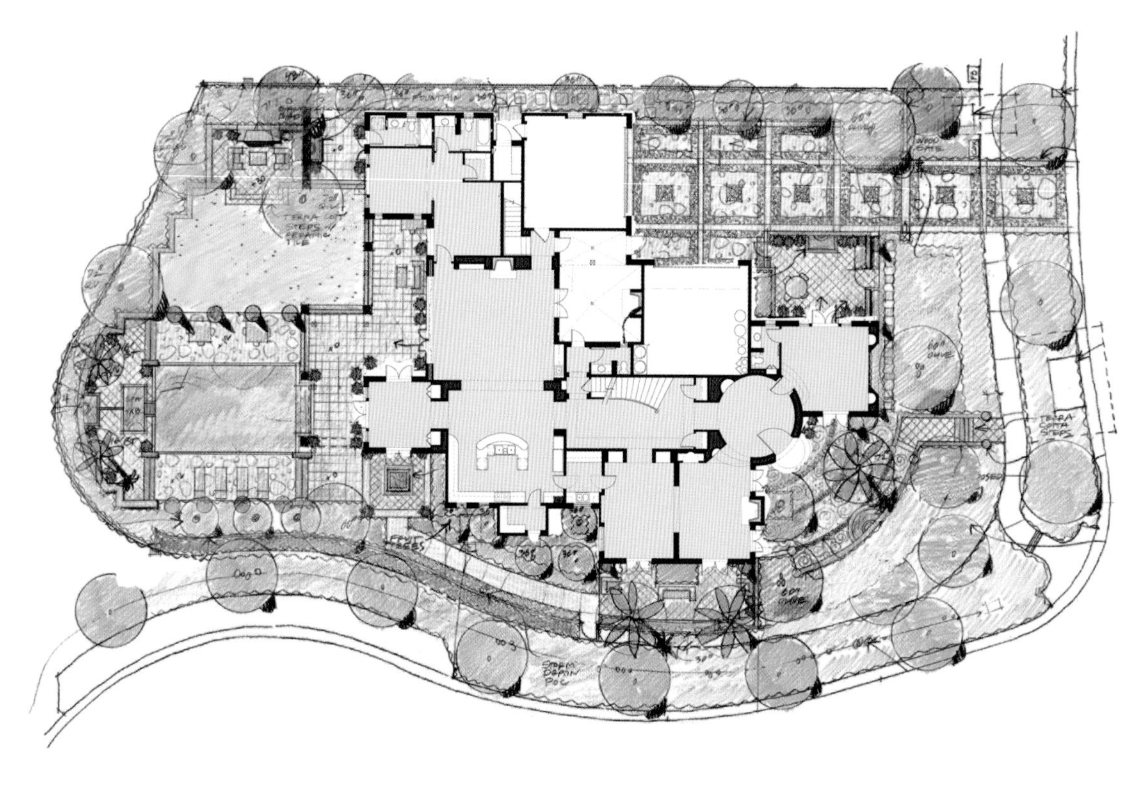

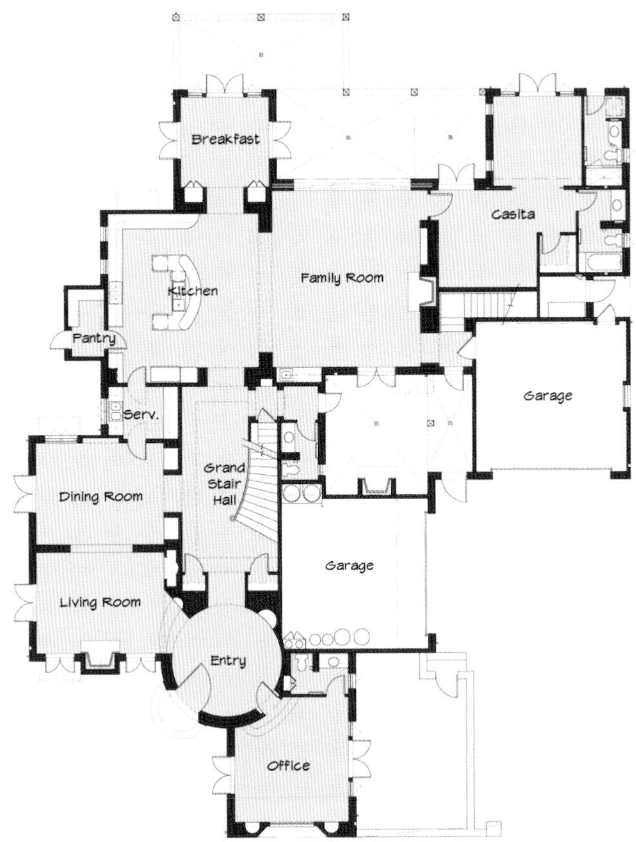
Breakfast
Casita
Kitchen
Family Room
Pantry
Garage
Serv.
Grand Stair Hall
Dining Room
Garage
Living Room
Entry
Office

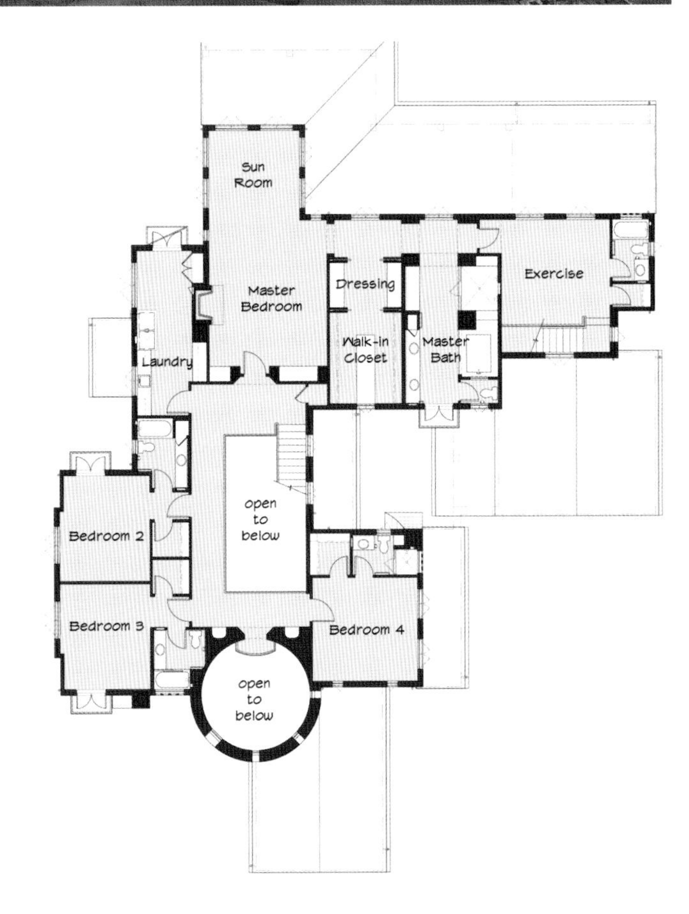
Sun Room
Master Bedroom
Dressing
Exercise
Laundry
Walk-in Closet
Master Bath
open to below
Bedroom 2
Bedroom 3
Bedroom 4
open to below

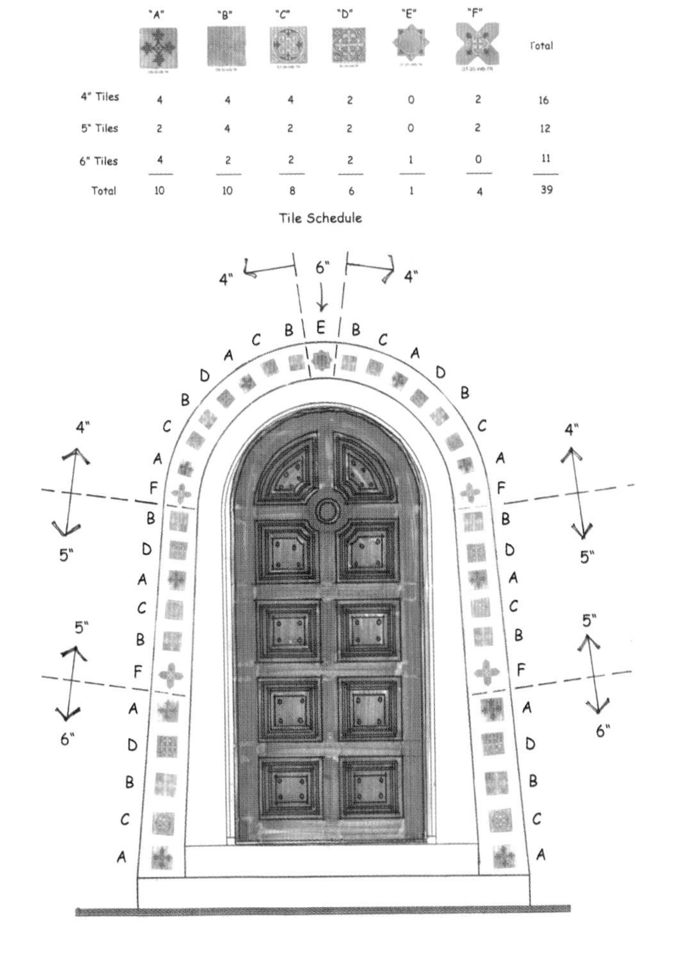

	"A"	"B"	"C"	"D"	"E"	"F"	Total
4" Tiles	4	4	4	2	0	2	16
5" Tiles	2	4	2	2	0	2	12
6" Tiles	4	2	2	2	1	0	11
Total	10	10	8	6	1	4	39

Tile Schedule